The Institute of Biology
Studies in Biology No. 64

Diseases in Crops

B. E. J. Wheeler
Ph.D.
Reader in Plant Pathology, Department of Botany,
Imperial College of Science & Technology,
University of London

Edward Arnold

First published 1976
by Edward Arnold (Publishers) Limited
25 Hill Street, London W1X 8LL

Board edition ISBN: 0 7131 2552 7
Paper edition ISBN: 0 7131 2553 5

Printed in Great Britain by
The Camelot Press Ltd, Southampton

General Preface to the Series

It is no longer possible for one textbook to cover the whole field of Biology and to remain sufficiently up to date. At the same time teachers and students at school, college or university need to keep abreast of recent trends and know where the most significant developments are taking place.

To meet the need for this progressive approach the Institute of Biology has for some years sponsored this series of booklets dealing with subjects specially selected by a panel of editors. The enthusiastic acceptance of the series by teachers and students at school, college and university shows the usefulness of the books in providing a clear and up-to-date coverage of topics, particularly in areas of research and changing views.

Among features of the series are the attention given to methods, the inclusion of a selected list of books for further reading and, wherever possible, suggestions for practical work.

Readers' comments will be welcomed by the author or the Education Officer of the Institute.

1976

The Institute of Biology,
41 Queens Gate,
London, SW7 5HU

Preface

It has been estimated that in his relatively short history man has used over 3000 species of plants for food and that at least 150 of these have entered world commerce. Even today the world's population relies especially on some fifteen species for food; these are the cereals—rice, wheat, maize, sorghum and barley; the sugar plants—sugar cane and sugar beet; the 'root' crops—potato, sweet potato and cassava; the legumes—common bean, soy-bean and peanut and the tree crops—coconut and banana. Diseases of these crops have wide-ranging economic and social consequences. Other diseases may also be serious because they affect crops which are locally important. Man attempts to minimize these adverse effects by various methods of disease control but to apply these intelligently requires some basic knowledge of how diseases develop in crops. This booklet looks at plant diseases from this essentially practical viewpoint and attempts to outline some of the more important developments in epidemiology and control of plant diseases over the past two decades.

London, 1976

B. E. J. W.

Contents

1 The Components of Disease

Several definitions of disease in plants have been proposed. Some examples are: 'a continuous impairment of metabolism', 'a series of harmful physiological processes caused by continuous irritation of the plant by a primary agent', 'a harmful deviation from the normal functioning of physiological processes'. These definitions imply collectively that disease is abnormal, injurious, physiological in nature and also a process and that it results from the interaction of some agent and the plant.

1.1 Pathogens

The agents of disease can collectively be termed pathogens. Purists may argue that this term should be reserved for agents which are themselves living but there is no generally-acceptable term for non-living, disease-inducing agents so pathogen is here applied to the entire range. The agents can be divided into eight groups: fungi, bacteria, viruses, mycoplasmas, nematodes, some other animals such as insects and mites, a few flowering plants such as dodder (*Cuscuta*), broomrape (*Orobanche*) and witchweed (*Striga*) and a mixed group (often described as 'non parasitic agents'), which includes mineral deficiencies and excesses and atmospheric pollutants such as sulphur dioxide. The fungi, bacteria and viruses collectively cause most of the known plant diseases and our examples will be restricted to these.

Fungi are plants which lack chlorophyll. Most of them are composed of threadlike filaments (hyphae) which are aggregated into a branched system (mycelium) from special parts of which spore-producing structures are formed, whose diversity and complexity provide the bases for classification. The vegetative body (thallus) of some fungi, however, is amoeboid and often in these instances the entire thallus is involved in the reproductive processes.

Bacteria also lack chlorophyll but basically are single cells which reproduce by division into two (binary fission). All plant pathogenic bacteria are rods, of general size 0.5 μm wide and 1–3 μm long and some have flagellae and are thus motile.

Viruses are nucleoproteins of various shapes; some are flexuous threads, others are rigid rods, spheroids or polyhedra. All are too small to be seen individually except in the electron microscope. Basically a virus particle (virion) consists of a central core of nucleic acid which is surrounded by a protein coat (capsid). Most plant viruses contain RNA (ribonucleic acid) but at least one, cauliflower mosaic virus, contains DNA (deoxyribo-nucleic acid). In plants the virus nucleic acid appears to re-direct the metabolism of the infected cell to the synthesis of more virus particles.

No mention of plant viruses is now complete without some comment on mycoplasmas. Organisms resembling species of *Mycoplasma* (the pleuropneumonia-like organisms) were first discovered in plants by Japanese scientists in 1967 and it is now clear that they cause certain diseases previously attributed to viruses. They differ from viruses in that they exist in varying forms (pleomorphic), they have a cell membrane and are particularly sensitive to certain antibiotics (the tetracyclines). Some *Mycoplasma* spp. from animals can be grown apart from their hosts in cell-free media and similar claims have been made for some mycoplasma-like organisms from plants.

1.2 Hosts

Plants which harbour or support the activities of pathogens are called hosts or host-plants. The pathogens for at least some stage obtain food from them. In this respect the pathogens are also parasites and although here we are concerned more with the changes in the host's physiology which ultimately result in damage to or death of cells, nevertheless the nature of the parasitism impinges directly on this. Thus some parasites such as the fungus *Pythium ultimum* rapidly kill the host tissues and succeed as parasites mainly because they are able to attack young seedlings of many species. By contrast other fungi, such as the smut fungus, *Ustilago nuda* on wheat and barley, succeed as parasites because they induce few adverse changes in the host throughout the growing season. Only at flowering does the pathogenicity of *U. nuda* become apparent in the replacement of the grain with fungus spores (Fig. 1–1).

As these two examples also indicate, many plant species may serve as hosts for some pathogens whereas in other instances either few are suitable or else only one kind of plant serves as host for a particular pathogen. The growing of certain plants as crops by man has had two important effects on fungal, bacteria and viral pathogens. The selection and breeding of particular types (cultivars) of a plant species, to satisfy commercial requirements such as high yield, have led to the development of strains specialized in their ability to attack these cultivars. The monoculture of these cultivars then ensures a uniform population of hosts within which these strains can develop.

1.3 The environment

Disease results from the interaction of a pathogen with its host but the intensity and extent of this interaction is markedly affected by the

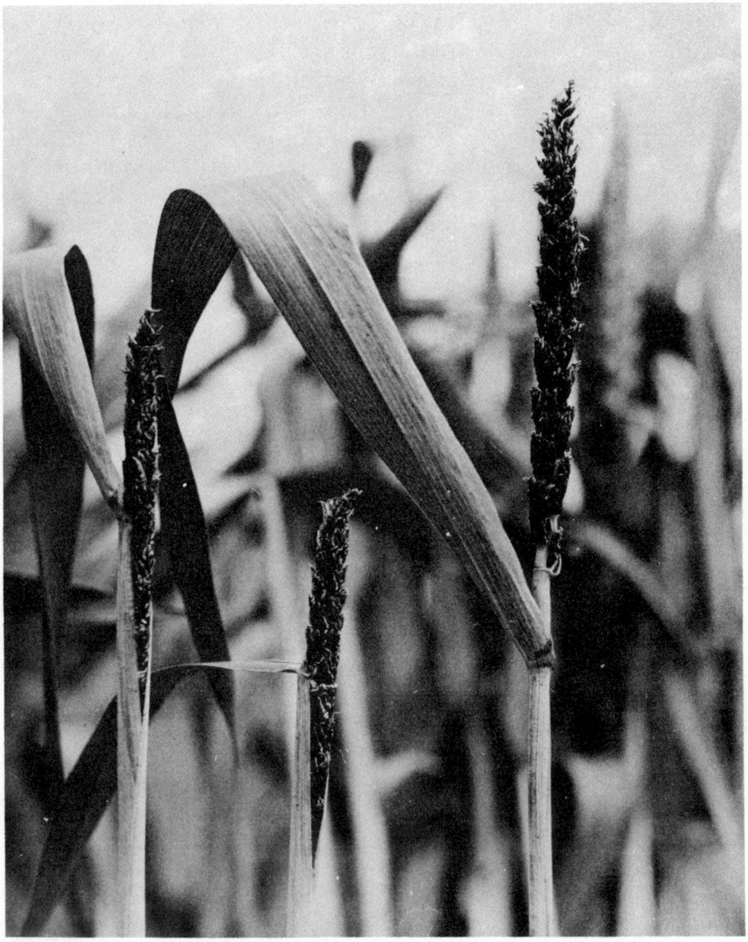

Fig. 1–1 Loose smut of wheat. Black spore-masses of the fungus *Ustilago nuda* replace the flowers when the head emerges. (Photograph: the National Institute of Agricultural Botany.)

environment and any consideration of disease in crops involves the 'disease triangle':

$$
\begin{array}{ccc}
\text{Pathogen} & \rightleftharpoons & \text{Host} \\
\nwarrow\!\searrow & & \swarrow\!\nearrow \\
& \text{Environment} &
\end{array}
$$

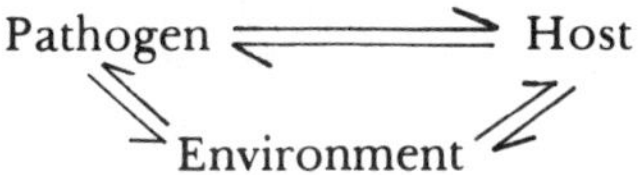

That environmental factors such as temperature, rain, air humidity, duration and intensity of light affect the growth of hosts is fairly readily accepted by most who can observe the seasonal changes in plants in parks and gardens. The effects on pathogens are less obvious but some can be determined in relatively simple experiments especially on events which precede the penetration of the host—for example, the effect of temperature or humidity on the germination of fungal spores. The effects of the host on the environment are perhaps even less obvious but may be considerable as a comparison of the humidity within a developing potato crop with that outside it indicates (Fig. 1–2). Clearly as the potato shoots grow and foliage meets across adjacent rows a humid atmosphere is

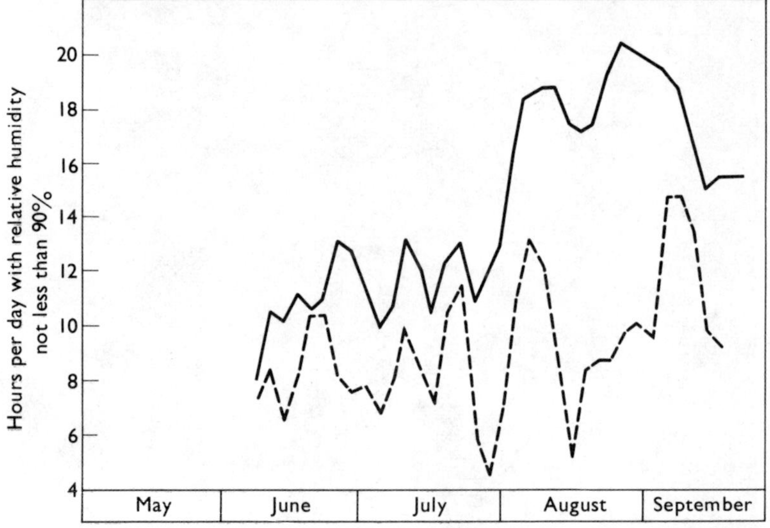

Fig. 1–2 A comparison of the number of hours per day with relative humidity not less than 90% within a crop of potatoes (——) and outside it (– – – –). (Based on HIRST, J. M. (1958). *Outlook on Agriculture* **2**, 16–26.)

retained under the canopy for much longer periods following rain. This favours the development of blight caused by the fungus *Phytophthora infestans*. Pathogens can also affect the humidity within the host canopy if they cause rapid defoliation.

Factors which determine micro-climate have a profound influence on pathogens attacking shoots but have less effect on root pathogens. These, however, are influenced by the many micro-organisms which abound in the region near the root (rhizosphere) to an extent which would partly justify the substitution of 'soil microbial population' for 'environment' viz.:

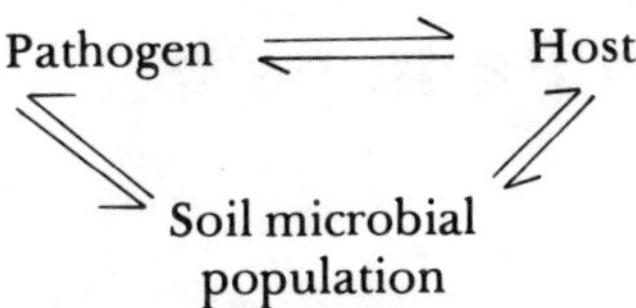

Root exudates, which largely account for this increase in general microbial activity, are also important in relation to the activity of some fungal pathogens: they stimulate germination of some resting spores, attract some motile spores to roots and influence the growth of some fungi at the root surface.

Having thus emphasized the main difference between the shoot and root as a milieu for the activity of plant pathogens the balance must be redressed by stating that the shoot surface also supports its own microflora though the effects on pathogens appear less marked. It should also be clearly understood that root pathogens and hosts, too, are influenced by temperature, moisture, pH and nutrient status of the soil.

2 Effects of Pathogens

2.1 Effects on the plant's physiological processes

The activities of a pathogen within its host result in changes which are visible either to the naked eye or can be observed by microscopy. These are the symptoms of the disease. To the trained plant pathologist symptoms have some diagnostic value. As well as providing a basis on which to measure disease within a crop they also indicate the main physiological functions of the plant which are affected by the pathogen. In the following brief review the activities of pathogens are considered in relation to seven physiological processes of plants under headings which also indicate the groups of diseases which are involved. This is a system suggested by an American plant pathologist, Dr. George McNew, which has been further developed in a text by ROBERTS and BOOTHROYD (see Further Reading).

2.1.1 *Effects on breakdown and utilization of stored food (damping-off and seedling blights)*

Crops are established by sowing seed or planting vegetatively-propagated material such as tubers, corms and stem cuttings. The basic feature of all such propagative material is that it is rich in food reserves. Given suitable conditions these materials are hydrolyzed into simple sugars which are then utilized in the early growth of the young plant. In some situations, however, these sugars are utilized by soil-inhabiting fungi with the result that the seed (or propagative material) rots before it can produce a seedling (pre-emergence damping-off) or else rotting of the very young seedling at the soil level results in its collapse and death (post-emergence damping-off) (Fig. 2–1). The pre-emergence damping-off of peas by the fungus *Pythium ultimum* is one example. The fungus is present in soil either as a mycelium living saprophytically on decaying plant remains on which relatively short-lived spore-bearing structures (sporangia) are formed or as thick-walled spores called oospores. The exudation of sugars following imbibition of water by the pea seed probably stimulates growth of the fungus. Indeed, there are indications that the increase in the disease with increasing soil moisture reflects the levels of sugars exuded. The fungus itself obtains further nutrients through its ability to produce extracellular enzymes which degrade the pectates of the middle lamellae, cause the host cells to separate and lead to abnormal leakage of materials from the host cytoplasm. Given suitable conditions of soil temperature and moisture the invasion of a seed can be extremely rapid, the tissues die within 30 hours and there is little evidence

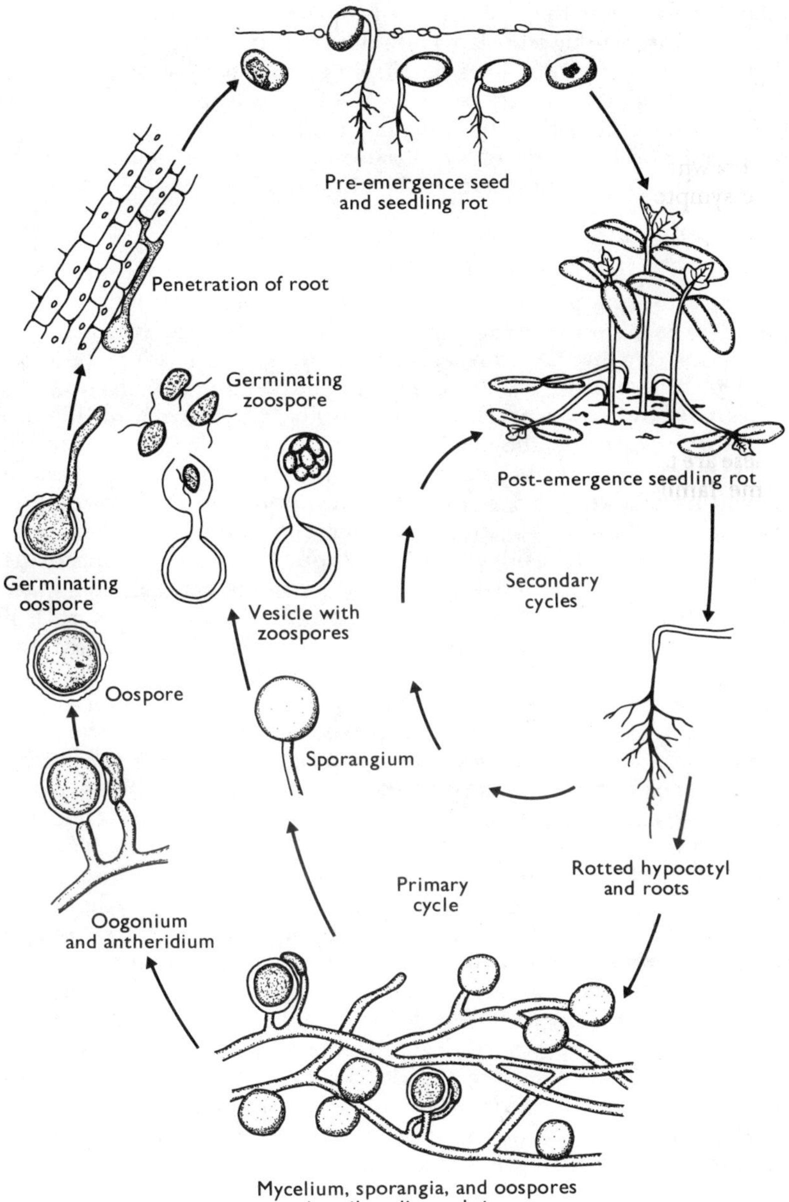

Fig. 2–1 *Pythium*-induced damping-off (From ROBERTS, D. A. and BOOTHROYD C. W. (1972).)

of any resistance by the host. Generally, however, the ability of fungi like *Pythium ultimum* to invade plants is limited to germinating seeds and very young seedlings. As seedlings develop they rapidly become resistant to attack. Indeed, the severity of damping-off reflects the balance between the growth of the pathogen on the one hand and the maturation of the host on the other and the effects of various environmental factors on the disease are most readily interpreted in these terms.

2.1.2 *Effects on absorption of water and minerals (root rots)*

The seed initially imbibes water from the soil but once the seedling root develops absorption of both water and nutrients occurs via its root hairs. Although in older plants much of the root system becomes lignified, absorption remains the primary function of the young 'feeding' roots and rootlets. The feature of these is that the central vascular core is surrounded by a cortex of relatively thin-walled parenchyma and it is this cortex which certain pathogens invade and rot and so impair absorption. The symptoms which they induce range progressively from a check in growth to yellowing and wilting of leaves, premature defoliation and eventually, collapse and death of the shoot. The rapidity with which these symptoms develop depends on the rate of rotting and on the ability of the host both to replace dead roots by new ones and to withstand the water shortage which results from the loss of roots. Generally root rots of annual plants result in the rapid development of symptoms whereas the rotting of the fibrous roots of a mature tree leads to a slow decline and dieback of the crown over several years. The pathogens involved are mainly fungi. Some are like *Pythium ultimum* in that they have a wide host range. Two examples are *Phytophthora cinnamomi* and *Armillariella mellea.* These are taxonomically quite different fungi. *P. cinnamomi* belongs to the same family (Pythiaceae) as *Pythium* but *Armillariella* is an agaric and produces its spores in the typical 'toadstool' fructification. *P. cinnamomi* has caused widespread destruction of eucalypts in Australia (the so-called 'Jarrah dieback'), is an important pathogen of avocado pear in the U.S.A. and also causes root-rot of a wide-range of ornamentals. It has become increasingly important in Britain recently especially on container-grown ornamentals for the 'garden centre' market. *Armillariella mellea* (the honey fungus) is a common pathogen in forests and gardens. It often forms aggregations of hyphae called rhizomorphs which grow through soil and subsequently over the surface of roots, penetration of these roots taking place some distance behind the limit of rhizomorph advance.

Other root-rotting fungi have a much more restricted host range. *Gaeumannomyces graminis*, for example, is restricted to the Graminae (grasses) and is most important on wheat and barley on which it causes the disease, 'take-all' (Fig. 2–2). It grows over the roots in a similar way to *A. mellea*, though as hyphae rather than rhizomorphs. Penetration similarly occurs some way behind the limit of mycelial spread. Infected cereals

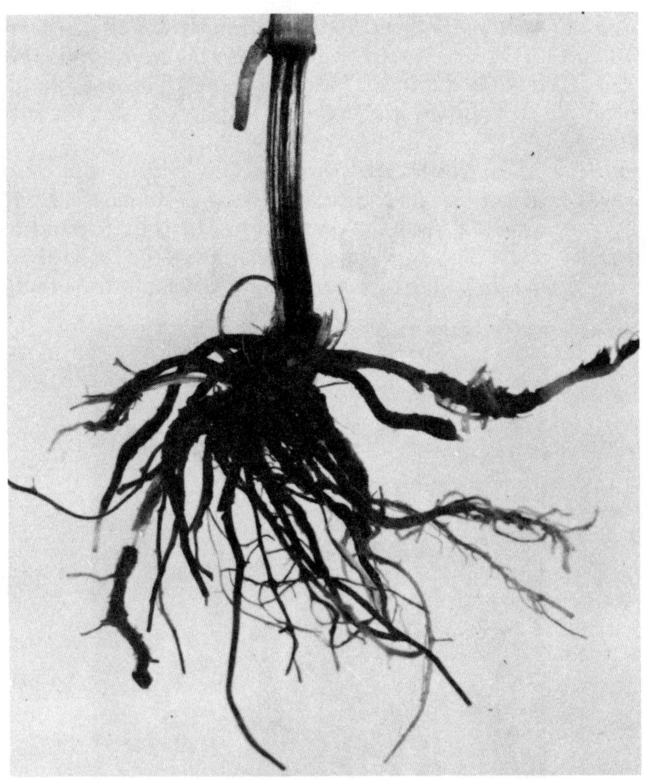

Fig. 2–2 A much-reduced and blackened root system of a wheat plant infected by the take-all fungus. *Gaeumannomyces graminis*. (Photograph: Rothamsted Experimental Station.)

usually bear little grain in their ears which thus have a bleached appearance and are often called 'whiteheads'.

2.1.3 Effects on water conduction (vascular wilts)

After absorption by the roots water and nutrients are conducted through the vessels and tracheids of the vascular system. It is a feature of some pathogens that once they have entered the plants they develop almost exclusively in this tissue. Some enter via the roots. Examples are *Verticillium albo-atrum*, a fungus which attacks tomato, hops, lucerne and many other hosts, *Fusarium oxysporum*, a fungus which exists in specialized forms able to attack particular crops like tomato, peas, cotton and banana, and the bacterium *Pseudomonas solanacearum* which is especially important on potato, tomato, eggplant, tobacco, groundnut and banana.

Other vascular-wilt pathogens, like the fungus *Ceratocystis ulmi* which causes Dutch Elm disease, are introduced into the stems of their hosts by insects. Once introduced into the vascular tissue all these pathogens may become rapidly distributed and, equally important, so may substances produced by them.

The effects on the hosts are complex and despite the description 'vascular wilt', wilting (i.e. flaccidity of leaves and stems) if it occurs at all is usually the last in a series of symptoms. In tomatoes attacked by *Verticillium albo-atrum* (Fig. 2–3). first the petioles of the lower leaves bend downwards so that the angle between each petiole and the stem becomes

Fig. 2–3 Wilt of tomato caused by the fungus *Verticillium albo-atrum*. The petioles of the lower leaves bend downwards (epinasty) and the leaves wilt progressively from the base of the plant.

unusually obtuse. The leaves then yellow progressively from the terminal leaflet. Some of these affected leaflets wilt during the day and then recover at night but this is followed by permanent wilting and desiccation and eventually the plant dies. Sometimes before this happens adventitious roots appear at the base of the stem and browning of the vascular region can often be observed if stems are cut. In other 'wilts' there may be additional symptoms, or one or two symptoms are more obvious than the rest. For example, in elms invaded with *Ceratocystis ulmi* yellowing and desiccation of leaves predominate and the tips of infected shoots often curl into a characteristic 'shepherd's crook'.

The physiological changes which result in these symptoms continue to be the subject of much speculation and research. In some instances the physical presence of the pathogen itself in the vessels may reduce the water flow but probably not enough to account alone for desiccation and wilting. It is more likely that this results from a complex interaction of pathogen and host involving the production of enzymes and extracellular polysaccharides by the pathogen, and the production by the host of gums and of growths into the vessel from the adjacent xylem parenchyma called tyloses (Fig. 2–4). The browning of the tissues suggests an involvement of phenol-oxidizing enzymes and changes in the petiole angle and production of adventitious roots suggest that growth regulators are also involved.

2.1.4 Effects on meristematic activity (leaf curl, witches' broom, club-root, galls and cankers)

Growth of the leaves, stem and root of the developing plant is initiated by division of cells at meristems and is controlled by certain chemicals or growth substances such as indole-3-acetic acid. Some pathogens drastically alter amounts of these growth regulators because they themselves produce these substances within the host tissues, or they induce the host to produce abnormally-large amounts, or both. As a result more and/or larger cells than normal are formed which leads to characteristic malformations. In leaves, for example, an increase in cells either side of the mid-rib and a stimulation in growth of the parenchyma cells of the palisade layer and spongy parenchyma result in a curling and puckering called leaf curl. This may be induced in different hosts by quite dissimilar pathogens—for example in peach by the fungus *Taphrina deformans* (Fig. 2–5) and in tobacco by leaf curl virus. In shoots abnormal growth processes lead to thick, short, much-branched stems or witches' brooms such as those induced by the fungus *Crinipellis perniciosa* on *Theobroma cacao* (the source of manufactured cocoa and chocolate). In roots the swelling can produce a 'club root' as in the infection of crucifers by the fungus *Plasmodiophora brassicae* (Fig. 2–6).

In all these examples, the plant organ retains enough of its original form to be recognizable but in other instances pathogens induce localized growths of host tissue or galls which bear no resemblance to any organ of the normal plant. The bacterium *Agrobacterium tumefaciens* does this on a wide range of hosts giving the disease, crown gall.

In woody hosts some pathogens similarly stimulate meristematic activity in local areas that they have invaded, but the host response is limited to the production of tissue (callus tissue) to seal off the infected zone. In succeeding years the pathogen often encroaches further on the host tissues and evokes further responses. The result is a dead area

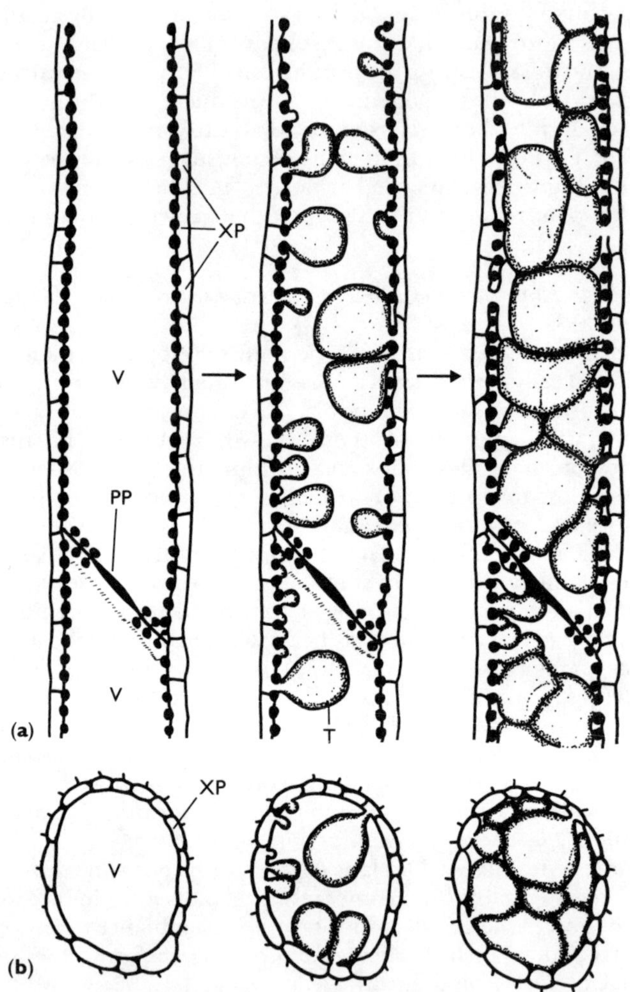

Fig. 2–4 Development of tyloses in xylem vessels. Longitudinal (**a**) and cross-section (**b**) views of healthy vessels (left), and of vessels with tyloses. Vessels on right are completely clogged with tyloses. PP=perforation plate; V=xylem vessel; XP=xylem parenchyma cell; T=tylosis. (From AGRIOS, G.N. (1969).)

Fig. 2–5 Peach leaf curl. A leaf infected with the fungus *Taphrina deformans* (*left*) contrasts with a healthy leaf (*right*). (Photograph by Ann Soley, Imperial College.)

surrounded by successive folds of callus tissues which is termed a canker. An example is the canker induced on apple by the fungus *Nectria galligena* (Fig. 2–7).

2.1.5 On photosynthesis (leafspots, mildews and rusts)

Growth of the host depends on its ability to synthesize sugars from carbon dioxide and water in the presence of light. Many pathogens affect this photosynthesis. Some do so because they kill the leaf tissue. The fungus *Phytophthora infestans* which causes potato blight is one example (Fig. 2–8); given suitable conditions it can destroy the shoots in a crop within 4 weeks. Generally it is considered that when 75 per cent of the foliage is killed no photosynthates are available for the developing tubers and they stop growing.

Other leaf-infecting pathogens affect photosynthesis in more subtle ways. This is true of the fungi causing downy mildews, powdery mildews and rusts and of certain viruses. For example, when barley leaves are inoculated with the powdery mildew fungus *Erysiphe graminis* (Fig. 2–9) there is initially an increase in photosynthesis but then after 4 days photosynthesis rapidly declines. The reduction obtained 7–10 days after inoculation does not always relate to the amount of the leaf surface covered by the pathogen: below 30 per cent coverage the reduction is more than can be explained just in terms of loss of leaf surface. Similar changes have been noted in leaves infected with rust; here reductions in photosynthesis appear to be linked especially with the sporulation of the fungus. Some of these affects can undoubtedly be attributed to loss of chlorophyll around pustules and indeed this can be seen clearly in some

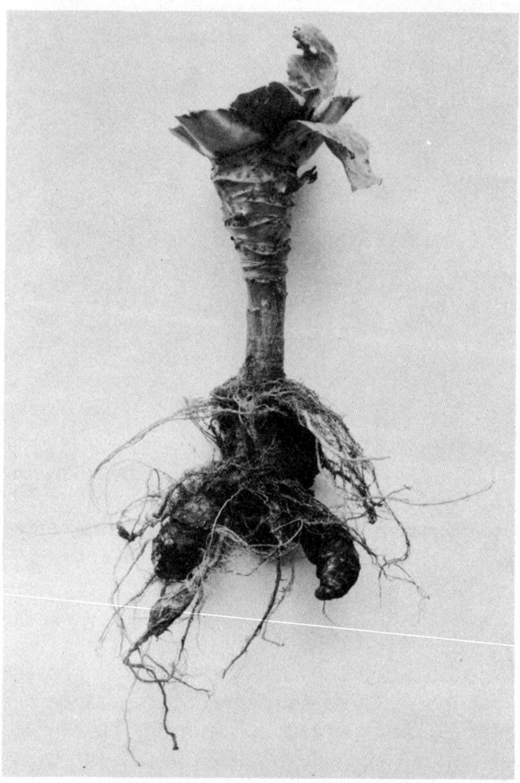

Fig. 2–6 Club root of cabbage. The swollen roots ('clubs') are invaded by the fungus *Plasmodiophora brassicae*. (Photograph by H. Devitt, Imperial College.)

rust infections (and those of other pathogens) where newly-erupted pustules are surrounded by haloes of abnormally pale host tissue. As the infected leaf senesces, however, there is frequently a redevelopment of chlorophyll in these areas so that the pustules appear in 'green islands' on the yellowing leaf.

Changes in chlorophyll are also suggested by the 'mosaics' and

Fig. 2–7 Apple canker caused by the fungus *Nectria galligena*. The young shoot is much swollen near the two cankers. (Photograph by Ann Soley, Imperial College.)

'mottles' produced in leaves by certain viruses (Fig. 2–10). In these there appear to be both a loss and a reduced rate of production of the chlorophyll pigments. Modern techniques for isolating intact chloroplasts from plants and studying their photosynthetic processes have greatly helped research into the induction of disease by these pathogens especially by those viruses, e.g., turnip yellow mosaic virus, which are adapted particularly to parasitize the chloroplast.

Fig. 2–8 Potato blight. The underside of a potato leaf showing two large lesions caused by the fungus *Phytophthora infestans*. (Photograph by H. Devitt, Imperial College.)

2.1.6 On translocation (diseases caused by viruses)

Sugars produced by photosynthesis are translocated to other parts of the plants, especially in the sieve tubes of the phloem. In the sense that they decrease the amounts of available photosynthates many pathogens also affect translocation. Similarly the observed accumulation of carbohydrates in the 'green islands' around rust and mildew pustules may be due to reduced translocation from the islands as well as increased photosynthesis within them. Other pathogens, notably some viruses and mycoplasmas, infect the phloem or they result in marked histological changes in the phloem, so it might reasonably be supposed that they specially affect translocation. Leaves of sugar beet infected with beet curly-top virus, for example, contain more glucose and sucrose than healthy ones mainly because translocation of these materials is about 100

Fig. 2–9 Growth of the fungus *Erysiphe graminis* on a barley leaf as seen in the scanning electron microscope. Spore-bearing stalks (conidiophores) arise from mycelium growing over the leaf surface. (Photograph by Pat Ready, Imperial College.)

times slower in infected plants. The virus multiplies in the phloem of leaves and induces several cellular changes. Cells immediately adjacent to the primary sieve tubes enlarge and subsequently die and cells near these divide but do not differentiate properly into new sieve tubes. Subsequently these abnormal sieve tubes and their companion cells also die.

2.1.7 Effects on food storage (post harvest diseases)

The developing seed, fruit or perennating organs such as tubers, corms and rhizomes are the ultimate destination of much translocated material and inevitably pathogens which affect synthesis and transport also markedly reduce the accumulation of these materials and so, obviously, in many crops affect yield. When the crops are harvested other pathogens often attack them in store; fruit and fleshy-storage organs rot and dried grain becomes mouldy. The pathogens are frequently those which

Fig. 2–10 Tobacco leaf infected with tobacco mosaic virus. (From WHEELER, B. E. J., 1969.)

develop not at all or only to a limited extent on the host in the field but usually they are present on at least some of the harvested material, possibly only as spores on the surface as with the fungus, *Fusarium solani* var. *coeruleum* which causes dry-rot of potato tubers. In most instances some damage to the tissue is necessary to initiate rotting. The lifting, grading and bulking of potatoes ensure some wounds through which the dry rot fusarium gains access to tubers. Once rotting is initiated its extent is largely governed by the conditions of the store and the physiological state of the storage organs. Moist, warm conditions favour rotting, hence the need for cool stores with adequate ventilation. Often the storage organs become more susceptible to rotting with increasing time in store particularly if this is associated with changes in dormancy, i.e. ability to produce young shoots as in potatoes.

Indeed, the disease problems of stored produce are rather different partly because the seed, fruit or other storage organ is a dormant structure and partly because the bulking of these tissues in store creates an environment quite different from that of the crop in the field. We shall not be considering this type of disease further in this booklet but it is well to bear in mind that even when a crop is harvested much may still be lost through disease.

2.2 Effects on yield

As indicated above, changes induced in plants by pathogens often cause damage and it might be supposed that in crops this results in a loss of yield. In most instances it is probably a fair assumption that there will be a yield somewhat below the potential for that crop in its particular environment. What cannot be assumed is that this loss in potential yield is enough to warrant controlling the disease and that such a measure can be justified in economic terms in most seasons. Some assessment of the intensity of the disease needs to be made in different years over several seasons and so the likely relationship between disease intensity and crop loss determined. Two examples will serve to illustrate some ways of tackling this problem.

The first relates to potato blight caused by the fungus *Phytophthora infestans*. As indicated above, tuber growth ceases when 75 per cent of the foliage is killed by blight. Clearly if this point is reached early in the growth of the potato crop, tubers will not develop fully and there will be a considerable loss in potential yield. Conversely if the 75% blight point is reached late in the season when tubers are nearing maturity, losses will be minimal. We could in fact simulate the effects of early and late attacks of blight by digging up samples of potato crops at different stages in their development. The upper graph of Fig. 2–11 shows such data. These were derived from sixty crops of Majestic potatoes grown at different centres in England and Wales during 1940–5 and from which there were fortnightly liftings. Probable losses of crops have been calculated from the

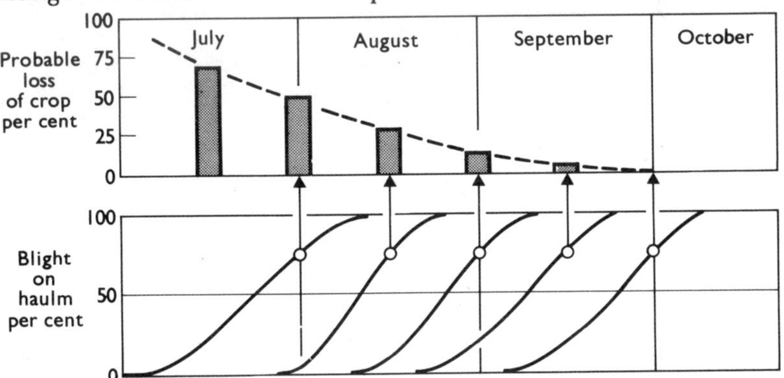

Fig. 2–11 Estimation of mean probable loss of crop from curves for progress of blight on the haulm (shoots). (From LARGE, E. C. (52) *Plant Pathology*, **1**, 108–17, by permission of the Controller, H.M.S.O.)

differences between the yields on these occasions and those for crops allowed to mature fully. The curves in the lower graph of Fig. 2–11 indicate the variations which occur in the development of blight in different regions of England in various seasons. Thus the far left-hand curve is typical of an early, severe attack on potatoes in the SW. and the curve on the extreme right a late attack on potatoes grown in northern England. Probable losses from blight are indicated by relating the 75% blight points to the cropping behaviour. Thus 75% blight by the end of July is likely to result in a 50% loss and 75% blight by the end of August in a 13% loss of yield. On this basis estimates of the losses due to blight have been derived for the four major potato-growing areas in England and Wales. Table 1 shows these estimates for the years 1947–56 and how they relate to the country as a whole, bearing in mind that the acreage of potatoes differs from region to region. For example, the climate in SW. England favours blight and it is often severe (mean loss 20%) but the crops there represent only 8% of the total grown and losses due to blight in the SW. amount to less than 2% of the national total.

Obviously these figures can only give a rough measure of the effects of blight. Other potato cultivars will crop differently and tubers may stop growing before there is 75% blight in the foliage. Nevertheless, the information provides a useful basis for appraising the likely benefits of control.

Our second example is concerned with powdery mildew of barley caused by the fungus *Erysiphe graminis*. For many years it was accepted that

Table 1 Estimated defoliation losses due to potato blight if all maincrops in England and Wales had been left unsprayed over a 10-year period, 1947–56. (From LARGE, E. C., 1958, *Plant Pathology*, 7, 39–48, by permission of the Controller, H.M.S.O.)

		Number of years in ten in which the 75% blight stage was reached by				*Mean loss per annum over the ten years*	
Zone	*Maincrop acreage % of total*	*Mid-Aug. 28% loss*	*End Aug. 13% loss*	*Mid-Sept. 4% loss*	*End Sept. no loss*	*to zone %*	*to whole country %*
South-west	8	5	5	0	0	20	1.7
Fens	13	4	2	1	3	14	1.8
Southern	51	0	5	2	3	7	3.7
Northern	28	0	0	5	5	2	0.6
England and Wales							7.8

some mildew developed in most crops of barley and it was generally assumed that losses were slight. The nature of the disease intensity–yield loss relationship was then investigated in a series of trials with spring-sown Proctor barley at nine centres throughout England over the years 1957–60. At each centre there were plots on which mildew was allowed to develop and others on which it was controlled by spraying every 2 weeks. The differences in yield between sprayed and unsprayed plots were then related to the amount of mildew at heading, i.e. the time when the ears had fully emerged but had not yet begun to ripen. The results indicated that yield loss was roughly equivalent to 2.5 times the square root of the mildew assessment (loss $= 2.5\sqrt{\text{mildew}}$). In these trials there was an overall mean of about 5% loss in grain. More recent general surveys of barley crops indicate annual losses due to mildew of 7–11% and powdery mildew is now rated as the most important foliar pathogen of barley in Britain. At least part of this change in attitude to the disease results from realistic appraisals of its effects on yield.

3 Development of Diseases in Crops

In the previous chapter we considered briefly some of the changes that occur within individual plants when they are invaded by pathogens. We now turn to the development of disease within a population of plants, a study which is termed epidemiology. Disease itself results from the interaction of a particular pathogen (fungus, bacterium, virus) with a particular cultivar, which is its host (p. 1) and there is a basic sequence of events by which this pathogen-host relationship is established. The pathogen and host must first come into contact with each other. Artificially we take, for example, a spore suspension of the fungus *Botrytis cinerea* and we place drops of this on the leaves of broad bean (*Vicia faba*) (DEVERALL, 1969, p. 7). We inoculate the plant. Under natural conditions disease establishment similarly starts with three things: the inoculum, e.g., fungal spores, the source of inoculum, e.g., plant debris on which the fungus has overwintered, and a means of establishing contact of inoculum with host, e.g., the dissemination of spores by wind and rain.

3.1 Inoculum and its source

We begin with the inoculum and its source. This involves looking at the ways in which plant pathogens perennate, that is how they survive between successive crops. This may be survival either between two plantings of an annual crop or between two periods of growth of a perennial crop. In the temperate zones perennation is often equated with overwintering, when the pathogen particularly has to withstand adverse temperatures. In the tropics, where cropping is associated with fairly regular periods of rain, perennation more frequently involves survival during a dry season and thus ability to withstand some desiccation. In general there are two principal ways in which plant pathogens perennate, in association with living host tissue and apart from the host (WHEELER, 1968). These will be considered in turn.

Pathogens that infect perennial plants have the advantage that the nature of their host facilitates their survival. Some which infect herbaceous perennials remain dormant in the perennating organs and then colonize the new tissues which are formed in the following year. The fungus *Pseudoperonospora humuli* which causes downy mildew of the hop is one example. It survives as mycelium within infected rootstocks and invades some of the dormant buds. The shoots which arise from these infected buds in the spring are shorter, thicker and paler than healthy ones and they bear on their leaves fructifications of the fungus which provide spores for initiating further infections.

Pathogens on woody perennials often remain within local areas of tissue that they have colonized, particularly those which cause galls, witches' brooms and cankers. Then in the following year they produce fresh inoculum on the infected area. Thus in the spring, masses of *Erwinia amylovora*, the causal bacterium of fireblight, exude from some cankers ('hold-over' cankers) produced in the previous season on pears and apples. Similarly, the fungus *Nectria galligena* produces its spores on cankered apple shoots.

Yet other pathogens perennate in buds on shoots of their perennial hosts. During the growing season the mycelium of the powdery mildew fungus *Spaerotheca pannosa* becomes established within some young buds of the rose and remains there protected by the buds scales until the following year (Fig. 3–1). When these buds burst they give rise to severely mildewed shoots which provide the initial inoculum for further infections.

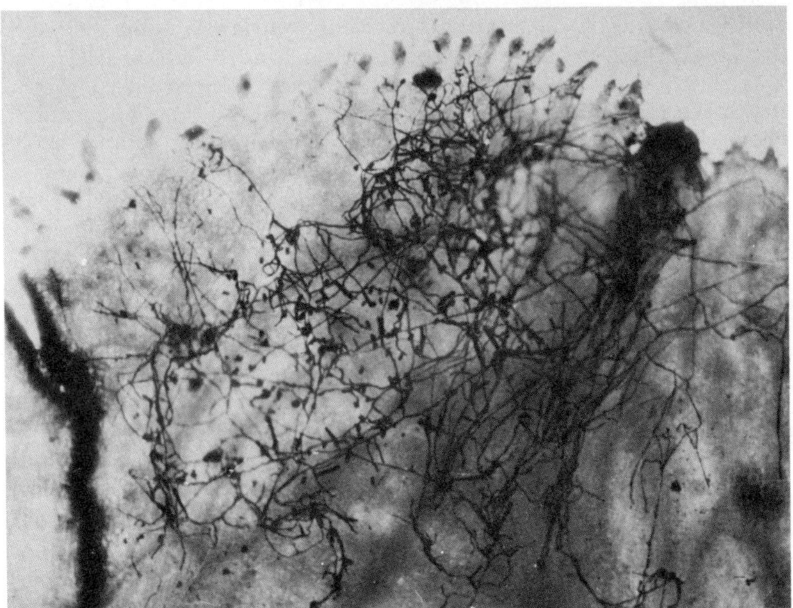

Fig. 3–1 Mycelium of the powdery mildew fungus *Sphaerotheca pannosa* in a bud scale of rose. (From PRICE, T. V. (1970). *Annals of Applied Biology*, **65**, 231–248.)

Some pathogens which infect annual crops also survive in association with their hosts because they infect or contaminate the planting material or seeds. The simplest way of ensuring survival of many viruses that infect potato is to plant 'seed' tubers derived from an infected plant. Similarly, potato blight can be initiated from tubers infected with the mycelium of *Phytophthora infestans*. Nor need these blighted tubers necessarily be planted to initiate the disease in a new crop. Sometimes, either at harvesting or when clearing out potato stores tubers are discarded. In the spring these sprout and some infected with *P. infestans* give rise to shoots in which the fungus grows and then sporulates and these serve as a source of inoculum for the new crop.

Survival of fungal pathogens on true seeds is very common and occurs in several ways. The causal fungus of loose smut disease of wheat and barley, *Ustilago nuda*, enters the scutellum and embryo at flowering without impairing the further development of the grain or its capacity to germinate. Indeed, the seed appears normal to the naked eye and at this stage the fungus can only be detected by a microscopic examination of the excised embryo. If such infected seed is planted in the following season, smutted plants result. Comparatively few fungi enter the embryo in this way but many invade the seed or fruit coat, remaining there either as resting mycelium like the loose-smut pathogen *Ustilago avenae* on oats, or as fruiting bodies such as those produced by *Septoria apiicola* on celery seed (Fig. 3–2). These can sometimes be seen with a hand-lens on celery seed soaked in water and appear as black dots. They are flask-shaped structures (pycnidia) partly embedded in the seed coat and contain needle-like spores which exude from an opening or ostiole. When infected seed is planted some spores wash down on to the cotyledons and initiate the destructive leaf spot caused by this pathogen. Other fungi survive only as spores contaminating the seed surface and they can be readily detected in washings of seed samples. The fungus, *Tilletia caries*, is carried on some samples of wheat seed in this way (Fig. 3–3). Infection takes place when the seed germinates; the fungus grows in the developing plant and eventually replaces much of the ear. However, the seed coat remains intact, usually until the grain is harvested, when affected grains split and spill out the black spores of the fungus which then contaminate the healthy seed.

Survival of plant pathogenic bacteria on seeds is also not uncommon and can, likewise, involve relatively deep-seated infections or superficial contamination of the seed. The perennation of *Pseudomonas phaseolicola* in the seed of dwarf bean (*Phaseolus vulgaris*) is of the former type. The bacterium causes both a leaf-spot characterized by a surrounding pale green halo, hence the name 'halo-blight', and also a mottling of leaves not unlike those of some virus infections. In recent years it has also been troublesome in Britain on the runner bean, *Phaseolus coccineus*, grown in home gardens and allotments.

In contrast, relatively few viruses survive in and are transmitted by true seeds—lettuce mosaic virus is one example—and usually only a low percentage of the seeds from an infected plant carry the virus.

When we look at survival apart from the living host we find that many

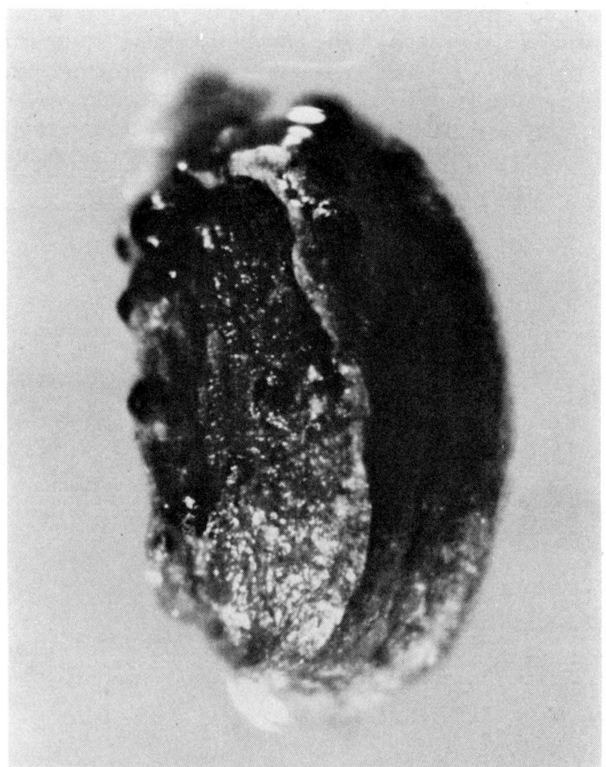

Fig. 3–2 Celery seed showing the black fruiting bodies (pycnidia) of the fungus *Septoria apiicola* embedded in the seed coat. (Photograph by A. Soley, Imperial College.)

pathogenic fungi form special structures capable of enduring adverse conditions (Fig. 3–4). The simplest of these are thick-walled spores like the resting spores of *Plasmodiophora brassicae* which are released into the soil from the decaying clubbed roots of infected brassicas. The oospores of *Pythium ultimum* and the thick-walled, vegetatively-produced spores (chlamydospores) of some *Fusarium* species, e.g., *Fusarium solani* var. *coeruleum*, also commonly have this function. Sometimes spores of this type germinate only after a period of dormancy, frequently coinciding with the winter season.

Other structures are more complex. For example with many pathogens belonging to the class of fungi called Ascomycetes the fruit body (ascocarp) itself overwinters. This is so for some powdery mildews such as *Sphaerotheca mors-uvae* on black currant. The fruit-bodies of this fungus are formed in late summer (from July to September in Britain) and appear as minute black spheres in brown patches of mycelium which contrast with the usual white growth of the powdery mildew. These fruit bodies (cleistothecia) each enclose a single sac-like structure (the ascus) containing eight spores and many remain on the fallen leaves throughout the winter. In contrast some fungi survive as mycelial aggregations called sclerotia. The ergot fungus, *Claviceps purpurea*, is one example. This

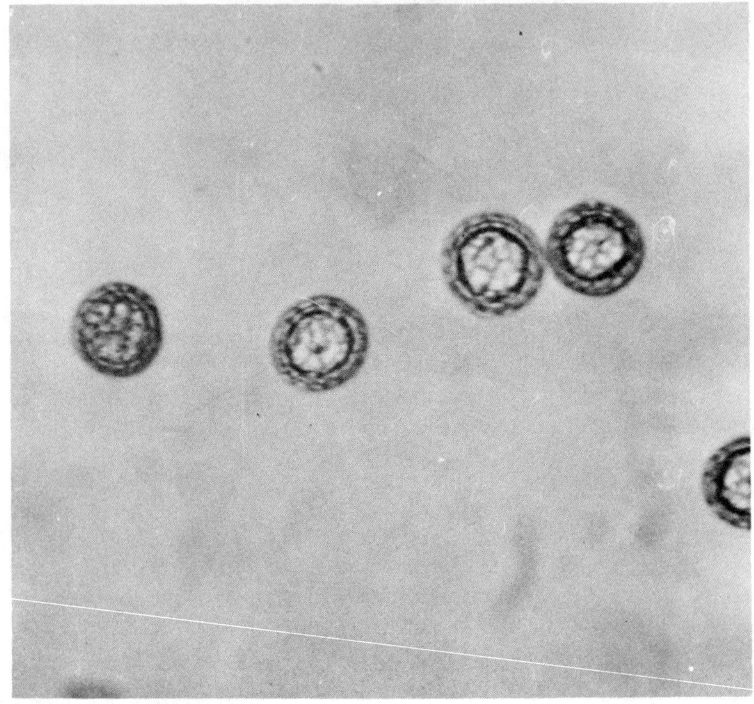

Fig. 3–3 Spores of the fungus *Tilletia caries* washed from a sample of wheat seed.

invades the flowers of some cereals and grasses and replaces the developing grain. The sclerotium which is formed falls to the ground and overwinters there. The white-rot fungus, *Sclerotium cepivorum*, on onions is another example. Its sclerotia also pass into the soil, in this instance from the rotting bulbs.

Other fungi do not survive by means of special structures but live as saprophytes in the decaying remains of their hosts. These they exploit in the face of increasing activity by other micro-organisms, some of which may be antagonistic. *Gaeumannomyces graminis*, the cause of take-all, colonizes the severed roots of wheat and barley in this way following harvesting. Provided there is enough nitrogen available, a viable

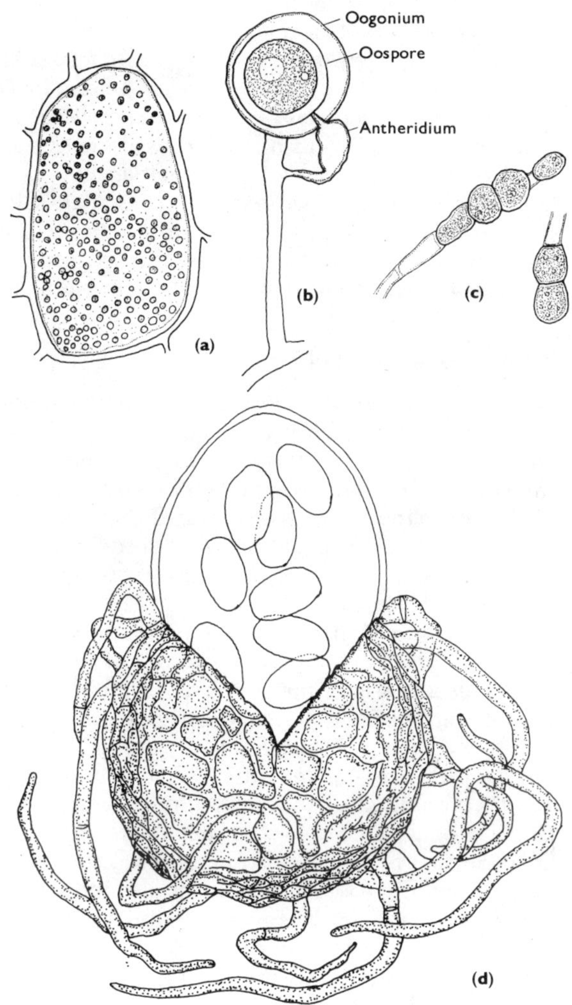

Fig. 3–4 Perennating structures of some pathogenic fungi
(**a**) resting spores of *Plasmodiophora brassicae* within an enlarged cell from an infected cabbage root
(**b**) oospore of *Pythium ultimum*, formed by the fusion of an 'egg-cell' (oogonium) and nuclei from an antheridium
(**c**) chlamydospores of *Fusarium solani* var. *coeruleum*
(**d**) cleistothecium of *Sphaerotheca mors-uvae*, split open to show the single large ascus with its eight ascospores (based on drawings of E. Punithalingam).

mycelium remains through to the next season and provides inoculum for the infection of a second cereal crop.

Compared with fungi, few plant pathogenic bacteria and viruses survive apart from their hosts. Some bacteria persist in crop debris particularly if this remains relatively dry. *Xanthomonas malvacearum* does so on trash from cotton crops in which it has caused the disease, bacterial blight. A few bacteria, e.g., *Agrobacterium tumefaciens*—the crown-gall organism, and *Pseudomonas solanacearum*—the vascular wilt pathogen, may also persist in soil. Amongst the viruses only tobacco mosaic virus is sufficiently resistant to survive in crop residues though some others persist in soil, probably in association with their nematode or fungal vectors.

3.2 Inoculation and dispersal of inoculum

The above brief review of perennation indicates that many fungal pathogens become separated from their hosts and they require some means of re-establishing contact. With *Gaeumannomyces graminis* this involves no more than the chance contact of a host-root with the debris of the previous crop which the fungus has colonized. In other instances there are more subtle mechanisms. If brassica seed, e.g., cabbage or cauliflower, is sown in soil harbouring the resting spores of *Plasmodiophora brassicae* the young seedlings become infected, which suggests that the developing roots influence spore germination in some way. Similarly the presence of onion stimulates the sclerotia of *Sclerotium cepivorum* to germinate. Many instances have now been recorded which indicate that root exudates influence dormant fungal propagules in this way. In only a few, however, is the response specific, in the sense that germination occurs only in the presence of plants which serve as hosts. The *Sclerotium cepivorum*/onion combination is one example; germination of the sclerotia is similarly stimulated only by related *Allium* spp., e.g., leeks and shallots, which are also hosts for this fungus. In most examples which have been investigated the roots of non-hosts also induce germination, for example those of the grass *Lolium perenne* stimulate germination of the resting spores of *Plasmodiophora brassicae*.

All these examples relate to soil-borne pathogens. For a powdery mildew fungus like *Sphaerotheca mors-uvae* which perennates on fallen leaves as a closed fruit body (cleistothecium), contact with the host is a two-stage process. First, the cleistothecium splits and the ascospores are forcibly ejected, then these are carried by wind to the blackcurrant leaves.

Initially for most pathogens which remain in association with their hosts these problems of establishing contact are automatically overcome but subsequently these, too, rely on mechanisms of dispersal to bring them in to contact with other host plants and so enlarge their distribution within the crop. Wind and rain figure largely in the spore dispersal of many fungal pathogens that infect the aerial parts and the monitoring of the concentrations of their spores in the air near crops is an important feature of modern epidemiological studies. For example in experiments at Rothamsted Experimental Station changes in the number of spores of the powdery mildew *Erysiphe graminis* above a crop of barley were determined from spore traps each consisting of a vertically-held brass rod, 25 × 0.5 cm diameter with a strip of cellophane, 25 mm wide, around the middle coated with a Vaseline-wax mixture to give a sticky surface. The seasonal pattern of spores caught agreed very well (Fig. 3–5) with that obtained by a much more sophisticated spore trap in which known volumes of air are sampled and spores impact on the sticky surface of a slowly-rotating drum. The use of such simple traps, however, also has its difficulties. The diameter of the vertical cylinder (Fig. 3–6a) is a critical feature which affects the efficiency of trapping particularly at different wind speeds and for accurate work size of cylinder needs to be determined in relation to the trapping of different kinds of spores.

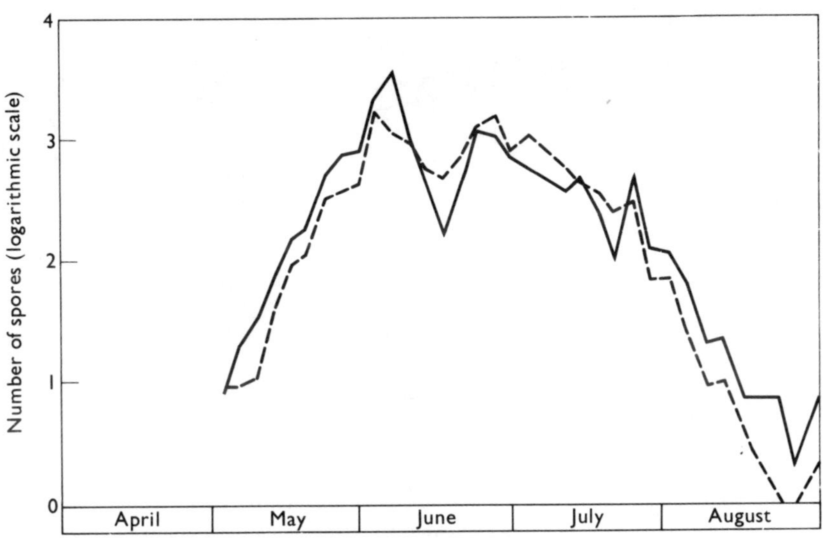

Fig. 3–5 Numbers of powdery mildew spores caught on a Burkard volumetric trap (. . . .) and on vertical sticky cylinders (———) placed within a crop of barley, 1971. (Based on JENKYN, J. F. (1974). *Annals of Applied Biology* **76**, 262.)

Some fungal spores appear to be dispersed almost exclusively by rain and for these different kinds of spore traps are required. These, too, can be quite simple in design such as strips of PVC guttering placed beneath bushes to collect water draining from leaves and channel it into a large container (Fig. 3–6b). The spores in these collections can be concentrated by centrifuging and resuspending them in a standard amount of water.

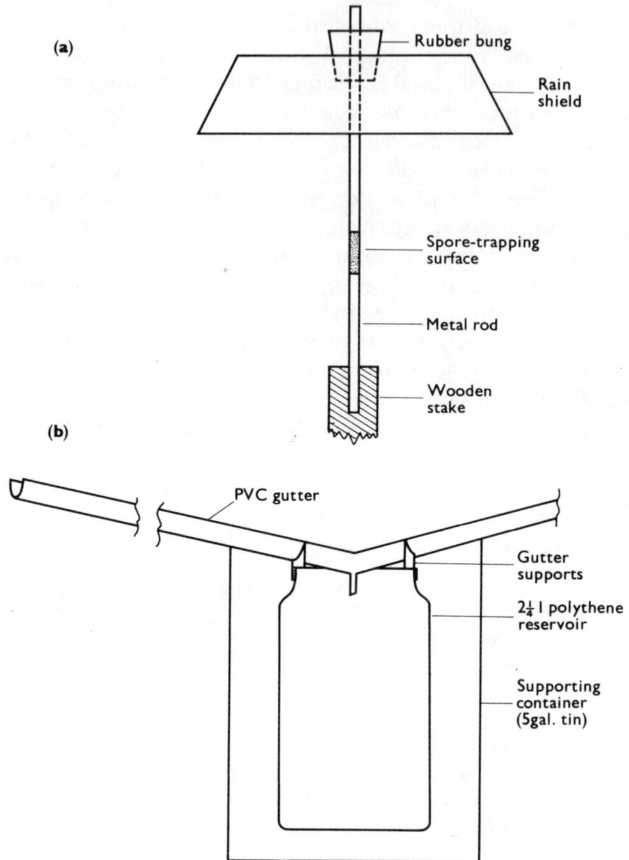

Fig. 3–6 Simple spore traps
(**a**) For wind-blown spores. The spore trapping surface consists of a strip of cellophane, 2·5 cm wide, coated with a Vaseline-paraffin wax mixture, wrapped around a metal rod (0·5 cm diameter, 25 cm long) and held to it by a narrow strip of Twinstick. This rod is held just above the crop level by inserting it in a wooden stake. Protection from the rain is afforded by a shield (a plastic 6 in flower pot saucer) held in place by two halves of a rubber bung. The trap can be kept in position for 3–4 days. The sticky tape is then removed, mounted in clear nail varnish and examined under the microscope. (Adapted from Jenkyn, J. F. (1974). *Annals of Applied Biology*, **76**, 258.)
(**b**) For rain-dispersed spores from small bushes or trees (from WALLER, J. M., (1972). *Annals of Applied Biology*, **71**, 3). The water draining from the leaves is caught in a PVC gutter placed below the canopy edge and collected in a polyethylene container. After rain, the water collected is measured, centrifuged and the deposit of spores re-suspended in a standard amount of water. The number of spores can then be measured in a haemocytometer. (See DEVERALL, B.J. (1969).)

They can then be counted in a haemocytometer (see DEVERALL, 1969, pp. 7–8). Rain splash also particularly favours the dispersal of bacteria; for example, drops impinging on exudates from fireblight cankers readily disperse *Erwinia amylovora*.

Viruses are not transmitted by these means but are dispersed mainly by insects and other animals, including man. The insects most often involved are aphids, leafhoppers and whiteflies. Basically, by feeding on an infected plant they acquire the virus and they transmit it when they move off to a fresh, healthy plant and feed there. There is, however, a good deal more to insect transmission of viruses than this simple statement suggests. Acquisition of the virus from an infected plant depends on factors such as the duration of feeding, whether it is continuous or not and whether the insect was starved or well fed when it alighted on the leaf. The virus may be carried externally on the mouthparts, e.g., on the aphid stylet (stylet-borne viruses), or it may be carried internally (circulative viruses); within the latter group the virus may multiply also in the insect vector (propagative virus) or it may not (non-propagative virus). Some viruses persist in their insect vectors for a long time, some only for a few minutes. These latter factors together with feeding time on the healthy plant influence greatly the development of a virus disease in a crop. So, too, do a complex of other factors such as the host range of the insect and its preference of species for food, the predators which attack it and the effects of the weather both on the host and insect movement.

Two groups of animals (other than insects) which are also important in virus spread are mites and nematodes (eelworms). One species of mite commonly encountered is *Phytoptus ribis* on blackcurrant; it transmits a virus which alters a leaf to a simple, less-indented pattern, a condition known as reversion. The association of viruses and certain species of nematodes, e.g., *Xiphinema index*, was first established in 1958 and has attracted much research since. One such virus is arabis mosaic which infects many plants including strawberry, raspberry, rhubarb and cherry. The nematodes concerned are widely distributed in soils and like the virus, have a wide host range. However, they do not move far in soil and spread of the virus by this means is limited. It is otherwise transmitted through the pollen and seed of infected plants and this accounts for its wider distribution.

Spread of viruses by man is best illustrated by tobacco mosaic virus (TMV). This can be transmitted simply by handling infected and healthy leaves in succession and spread in the field is effected largely by movement of workers or even machinery through the crop. The virus is so durable that some may persist in the cured leaf so that workers who smoke and then handle susceptible crops are likely to transmit the virus. TMV is often spread in glasshouse-grown crops of tomatoes in this way.

It would be wrong to associate insects and other animals only with virus dispersal. Several bacterial plant pathogens are dispersed by insects, most of them just by chance. A fly alighting on the bacterial ooze from a fireblight canker will carry *Erwinia amylovora* to the flowers that it then visits for nectar. A few bacteria have a much more intimate association with their insect vectors. *Pseudomonas savastanoi*, for example, lives in the gut and enters the egg of the olive fly, *Dacus oleae*, so that when this insect lays its eggs on the olive tree it also inoculates it with the bacterium. This induces small swellings on young twigs and leaves giving the disease name, olive knot.

Some fungi are also dispersed by insects. A notable example is the cause of Dutch Elm disease, *Ceratocystis ulmi*, which is spread by two species of *Scolytus*. The fungus sporulates freely in the galleries made by these beetles in the bark of infected trees and in which their pupae develop. When the new adult beetles emerge they carry the spores to young elm twigs on which they feed and so inoculate healthy trees.

3.3 Development of an epidemic

Dispersal of a pathogen by one or more of the mechanisms outlined above leads to the development of a disease within a crop to the point where many individuals are infected, that is an epidemic. In order to follow the progress of an epidemic we must first have a means of measuring disease. The simplest method is to count the number of infected plants in a known sample and so obtain percentage infection. This is best used in situations where presence or absence of infection is all that matters; for example, the plant is killed or it is infected with a virus and will thus remain so. The incidence of virus yellows in fields of sugar beet is assessed in this way from ten samples of one hundred plants at equal intervals along its diagonals. Infected plants are readily recognized by the crinkling and distortion of the youngest leaves and the yellowing, especially between the veins, of the older leaves.

With many diseases, individual plants are affected to different degrees and other assessments have to be used. Detailed accounts would be inappropriate here, but mention of two general methods will aid further discussion of epidemics. For some diseases descriptive keys have been evolved which aim to assess overall percentage infection within a field. That devised for potato late blight is shown in Table 2 and was the basis on which the disease was surveyed for the data of Table 1. Other diseases are assessed by means of standard diagrams which depict different levels of disease on a leaf as a percentage infection. Figure 3–7 shows a typical example; it attempts to simulate increasing amounts of the powdery mildew fungus (*Erysiphe graminis*) on cereal leaves. Diagrams like this are usually used to assess disease at particular stages of host growth such as heading (p. 21).

Table 2 Descriptive key for potato blight (ANON (1947), *Transactions of the British Mycological Society*, 31, p. 140).

Percentage blight	*Description*
0	Not seen on field
0.1	Only a few plants affected here and there; up to one or two spots in 12 yd. radius
1	Up to ten spots per plant, or general light spotting
5	About fifty spots per plant or up to one leaflet in ten attacked
25	Nearly every leaflet with lesions, plants still retaining normal form: field may smell of blight, but looks green although every plant affected
50	Every plant affected and about one-half of leaf area destroyed by blight: field looks green flecked with brown
75	About three-quarters of leaf area destroyed by blight: field looks neither predominantly brown nor green. In some varieties the youngest leaves escape infection so that green is more conspicuous than in varieties like King Edward, which commonly shows severe shoot infection.
95	Only a few leaves left green, but stems green
100	All leaves dead, stems dead or dying

Successive measurements based on any of these methods usually give a similar pattern illustrated by the blight curves in Fig. 2–11. Disease development is initially slow because there is relatively little inoculum, then it accelerates and finally slows again because now there is little healthy host tissue left for the pathogen to colonize.

In recent years there have been several attempts to 'fit' data of this type to various mathematical equations. The equation thus becomes a 'model' of the epidemic and the effects of various factors on the epidemic can be judged by assigning different values to the terms in the equation. One of the more useful models is that of VAN DER PLANK (1963). He suggests that at least in its early stages an epidemic can be described by the equation

$$x = x_0 e^{rt} \tag{3.1}$$

where x = the proportion of disease at any one time,
x_0 = the amount of initial inoculum,
r = average infection rate,
t = time during which infection has occurred.

The value e in the equation is the base of natural logarithms ($e = (1 + \frac{1}{n})^n$, where n is a very large number, $= 2.718 \ldots$). We are thus dealing here

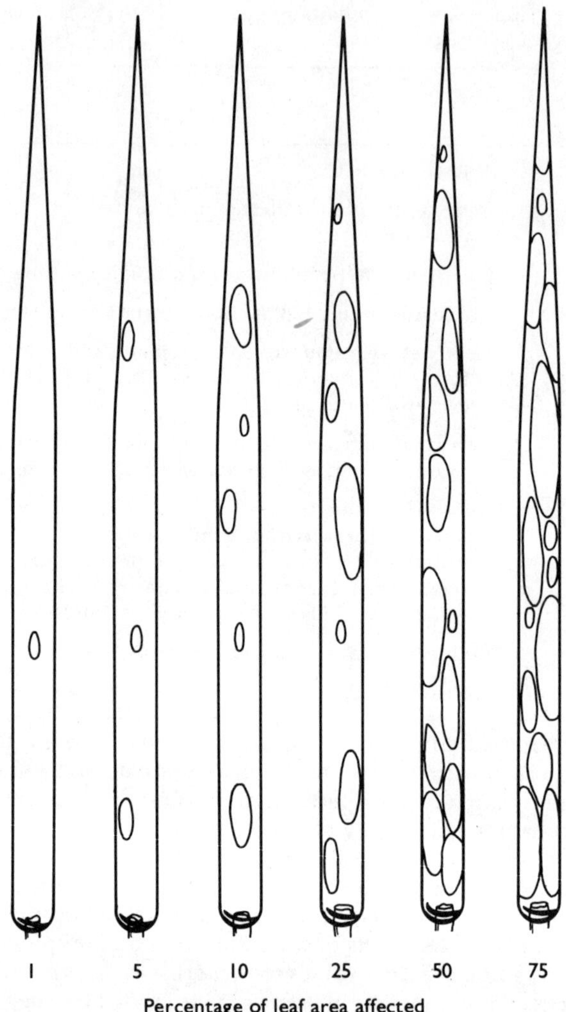

Fig. 3–7 Standard diagram for assessing barley powdery mildew. (Based, by permission, on the key used at the Ministry of Agriculture, Fisheries and Food, Plant Pathology Laboratory, Harpenden. Crown Copyright.)

with an exponential function and the basic assumption is that at any one moment the rate of disease increase is proportional to the amount of disease present at that moment. In the early stages of the epidemic when there is plenty of susceptible tissue for the pathogen to colonize this is likely to be true but obviously it is not at the end as the blight curves in Fig.

2–11 indicate. Indeed, as the epidemic progresses its rate is determined not only by the amount of disease present x, but by the proportion of susceptible tissue left $(1-x)$. The stage at which this point is reached probably varies with the pathogen-host combination. For potato blight it is considered, somewhat arbitrarily, to be reached when there is 5% infection, i.e. the proportion of disease x, is 0.05.

The item usually of most interest in equation (3.1) is r. It gives us an overall measure of the rate at which the epidemic is progressing and can be used to compare epidemics of the same disease in different localities and within different cultivars. It is derived by taking logarithms through equation (3.1) and transposing:

$$\log_e x = \log_e x_0 + rt \qquad (3.2)$$
$$rt = \log_e x - \log_e x_0$$
$$r = \frac{1}{t} \log_e \frac{x}{x_0}$$

Provided, therefore, we have an initial measurement of disease and a second some time afterwards we can calculate average rate of increase over that period of time. By convention the first measurement is x_1 (assessed at time t_1) and the second x_2 (assessed at time t_2). Average infection rate is then calculated as

$$r = \frac{1}{t_2 - t_1} \log_e \frac{x_2}{x_1} \qquad (3.3)$$

For example, in a particular trial on barley in 1972 there was 0.5% and 3.5% powdery mildew in the crop at 45 days and 55 days after planting respectively. So, $x_1 = 0.005$, $x_2 = 0.035$ and $t_2 - t_1 = 10$ days, therefore

$$r = \frac{1}{10} \log_e \frac{0.035}{0.005}$$
$$= 0.194 \text{ units of disease per day.}$$

We are here assuming that there is plenty of susceptible tissue, i.e. $x \leqslant 0.05$. With higher disease levels the formula for calculating r is more complicated because the factor $(1-x)$ is introduced:

$$r = \frac{1}{t_2 - t_1} \log_e \frac{x_2(1-x_1)}{x_1(1-x_2)} \qquad (3.4)$$

The rate of infection in equation (3.3) is sometimes called logarithmic infection rate and is denoted by l as a subscript r_l, to distinguish it from r in equation (3.4) which is then called 'apparent infection rate'. An alternative method of calculating rates of infection where there are several disease assessments is to plot $\log_e x$ against time (if x remains low throughout) or $\log_e [x/(1-x)]$ if disease levels go beyond about 5%. The slopes of the lines (which can be fitted mathematically) directly measure η or r respectively (Fig. 3–8).

It may well be that epidemics of some diseases are better described by different equations—this is an area of continuing investigation. Van der Plank's model has, however, greatly influenced current thought about epidemiology and control as the next chapter illustrates.

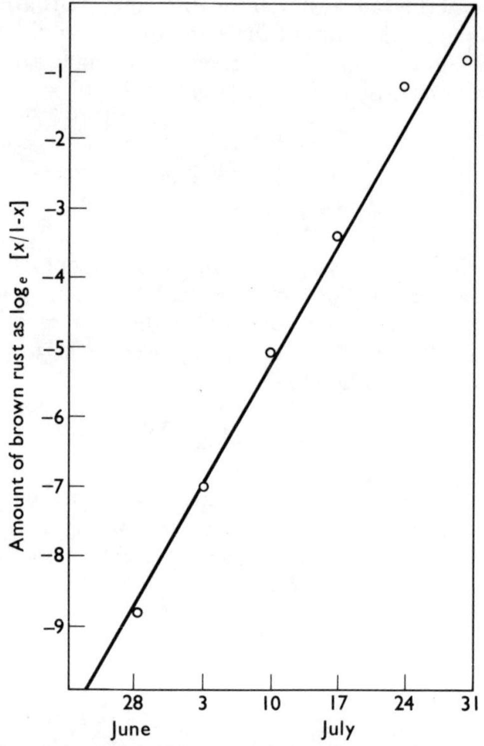

Fig. 3–8 Rate of infection of brown rust (caused by the fungus, *Puccinia hordei*) in a crop of barley, 1973. (Based on data supplied by ONURSAL, N. F.)

4 Plant Disease Control

Van der Plank suggests that the development of a plant disease in a crop can be described by the equation: $x = x_0 e^{rt}$ (p. 33). This immediately indicates several basic ways in which we might control disease, i.e. reduce the value of x. We could, for example, attempt to get rid of all the initial inoculum x_0, or at least reduce it. Alternatively we could attempt to lower the infection rate r, or reduce the time t, over which infection can occur. Sowing crops early and using early-maturing types effectively limits the period over which crops can be infected and these are useful ways of limiting disease but in practice most control methods relate to initial inoculum and/or infection rate and our discussion will be limited to these.

4.1 Excluding the inoculum

Some diseases are always with us. Blight, for example, occurs year after year in most areas of Britain and northern Europe where potatoes are grown. We say that it is endemic in these areas. Its intensity varies; the plant pathologist distinguishes 'blight years' when it is severe from 'non-blight years' when it is not, but some blight can usually be found in potato crops. It was not always so. Before 1840 there is little evidence that blight occurred in Europe but a few years later it was devastating potato crops. The Great Potato Famine in Ireland and its social consequences have passed into history but still make horrifying reading (see WOODHAM-SMITH, 1963, and LARGE, 1940). There is some speculation about how the fungus, *Phytophthora infestans*, was introduced, but it seems most likely that it was brought on potatoes from the Americas and that the more rapid passages by steamship facilitated the survival of the fungus.

Today most countries have regulations which aim to exclude pathogens not yet present in the territory. For example, Britain has regulations concerning the importation of potatoes to exclude, amongst other things, *Pseudomonas solanacearum* the cause of bacterial wilt and the West African territories which grow cacao have regulations to exclude *Crinipellis perniciosa*, the causal fungus of witches' broom. The enactment of this legislation is the responsibility of the territory concerned. However, such regulations inevitably affect trade and certainly require co-operation between territories if they are to be truly effective. These international aspects were recognized in 1951 by the drawing-up of an International Plant Protection Convention which formulated certain principles for co-operation in the field of plant protection. Since then organizations have been formed which deal specifically with problems common to states within certain regions such as the European Plant Protection Organization (EPPO).

Plant disease legislation may also be concerned with the movement of diseased material within a country. In the U.S.A. provision is made for controlling the movement of plants between states and even within a territory as small as Britain there are some restrictions on the movement of plant material. In some instances this is aimed at limiting a particularly virulent pathogen. For example, about 1930 a severe form of hop-wilt caused by the fungus *Verticillium albo-atrum* developed in parts of SE. England. It was so destructive that it acquired the name 'progressive wilt' to distinguish it from the more common form ('fluctuating wilt') which seldom killed the plant completely. To prevent the spread of this virulent strain of the fungus to other farms and especially to the hop growing areas of the south Midlands a special order (*Progressive Verticillium Wilt of Hops Order*) was promulgated which made it obligatory for farmers to report outbreaks of this type and prohibited them from selling planting material.

In other instances legislation aims to limit the sale of diseased plants in general by requiring that only certified planting stock of some crops are sold, that is stock which has been inspected and which meets certain prescribed standards. Potatoes for planting ('seed' potatoes) are so certified in most developed countries.

The principle of excluding inoculum is applied to individual farms and smallholdings by this practice of planting pathogen-free stock. The means by which this stock is obtained are varied and often ingenious. Planting material is sometimes produced in areas where certain pathogens are less common or are less likely to infect. In Britain seed potatoes are produced in Scotland and in the higher areas of northern England and Wales where conditions are less conducive to the build-up of aphids within the crops and so the risk of viral infections is less. Even so, there must be some nucleus of virus-free potatoes from which these stocks can be propagated. Two techniques have greatly helped the production of virus-free material not only of potatoes but also of many other crops. They are heat-treatment and meristem culture. Heat treatment is based on the discovery by an American plant pathologist, Dr. L. O. Kunkel, that small peach trees could be cured of virus yellows by keeping them at 35° C for 2–4 weeks. The practical application of this work, which was published in 1936, was long delayed. It was not until the early 1950s that Dr. Kassanis and his colleagues used similar techniques at Rothamsted Experimental Station to eliminate potato leaf roll virus from tubers. About this time they, and others, also examined the possibility of growing virus-free plants from excised meristems because it had been discovered that virus concentration was least near the growing points of infected plants. The technique involved cutting-off under aseptic conditions about 100 μm of tissue from the tips of sprouts from tubers and growing these on sterile media until they developed into small plantlets. These plantlets were then examined for virus by various methods and those found to be virus-free grown on further. Modifications of this technique are now widely used in the initial production of virus-free stock and often heat treatment and meristem culture are combined. Plants are first kept in 'hot-boxes', usually at 37–39° C and a high relative humidity for several weeks to encourage the production of new shoots and meristems are then excised from these.

Once a nucleus of virus-free material is obtained it is multiplied as indicated in conditions unfavourable for virus infection. With potatoes, tubers from virus-free plants are multiplied over 3–5 years. During this propagation, plots are examined visually for symptoms of disease and other, more sophisticated methods of detecting virus are also used. For example, leaflets are taken from plants, the sap is expressed from them and this is mixed with antiserum specific for certain potato viruses. The presence of one or more of the viruses in the plant sap is indicated by a precipitate. A similar test is used to detect virus in harvested tubers, a method called 'tuber-indexing', which can conveniently be carried out during the winter. In this instance sap is obtained from young shoots which develop when a tuber piece containing an 'eye' (i.e. bud) is sprouted in an insect-proof glasshouse. Tubers giving a positive test for virus and those from the same plants are discarded and further stocks propagated only from the remainder.

Heat treatments of a somewhat different type have been used to eliminate other pathogens from planting material, notably fungi from seeds. For example, immersion in warm water was used for many years to eliminate the smut fungus, *Ustilago nuda*, from wheat and barley seed. There were many variations in the length and temperature of the soak ranging from 10 min at just above 51° C to 5–6 hours at about 41° C. The method is basically simple but there are many practical problems with large quantities of seed. If the grain temperature exceeds the required level germination is impaired; if it does not reach it the fungus is not killed. Deep-seated pathogens in seeds like *U. nuda* have long been regarded as the most difficult to deal with but in recent years the situation has improved with the development of fungitoxic chemicals which are absorbed into seeds without impairing germination. One such material, 2,3-dihydro-6-methyl-5-carbamoyl-1, 4-oxathiin (carboxin) successfully eliminates *U. nuda* from wheat and barley seed.

In contrast chemicals have been used for many years to deal with fungi as spores contaminating the seed surface or as mycelium in the surface layers. The many different materials have ranged from simple compounds like copper sulphate and mercuric chloride in the early days of plant protection to more complex materials which are used today such as organic compounds of mercury and sulphur and the quinones. These materials are applied either as dusts or mixed with small amounts of liquid (slurries) or as concentrated solutions of the active ingredient. Regrettably many materials, especially the mercurials, have a high mammalian toxicity so not only is the preparation of the dressed seed hazardous but also its use.

4.2 Reducing the inoculum

Sometimes pathogens are introduced into new areas despite all attempts to exclude them. The problem is then to eradicate them, hopefully to remove the inoculum completely or at the very least to reduce it. There is a similar problem when pathogens are established in an area and perennate successfully.

4.2.1 Eradication of pathogens

There are now many examples of the sudden appearance of a disease within a territory and of a corresponding eradication scheme to deal with the pathogen. Relatively few of these schemes have been entirely successful but the elimination of citrus canker from the U.S.A. is an exception. *Xanthomonas citri* which causes the disease was probably first introduced into the Gulf states in 1910 on stock from Japan. At first citrus canker was not recognized as new, so some 4 years elapsed before any eradication scheme was put into operation. Then between 1914 and 1934 some 20 million trees were removed and destroyed within the citrus growing states and the point was reached when canker was no longer present in the commercial plantings. There remained, however, numerous abandoned groves and wild citrus where the disease could be found and further eradication programmes were mounted to deal with these. A review in 1964 indicated a complete elimination of the pathogen.

Other eradication schemes have been less successful. The felling of diseased elms and the destruction of elm logs did not stop the spread of *Ceratocystis ulmi*, the causal fungus of Dutch Elm disease, across the U.S.A. from Ohio where it was first found in the early 1930s. Nor did the destruction of pear trees with symptoms of fireblight prevent the further spread of the bacterium *Erwinia amylovora* following its introduction into the Kent orchards of England in 1958. There are several reasons for these apparent failures. The movement of elm logs and elm products westwards across the U.S.A. by railroad and the spread of the fungus by bark-boring beetles militated against the eradication of Dutch Elm disease. The widespread distribution of alternative hosts for *Erwinia amylovora*, the hawthorns (*Crataegus* spp.), and also the many small plantings in private gardens of the commercial hosts, apple and pear, similarly made the complete eradication of fireblight in England almost an impossibility. Nevertheless the schemes in both instances limited the spread of the disease and in this respect were of some value.

Physical removal of the entire diseased plant is not the only method of reducing inoculum. In some situations this can be achieved simply by removing the part of the plant that is diseased, e.g., a *Nectria* canker on an apple. In other situations much inoculum can be destroyed *in situ*. In seedbeds the inoculum of damping-off fungi, e.g., *Pythium ultimum*, can be killed by treatment of the soil with fumigants such as methyl bromide and chloropicrin or by sterilizing it with steam.

There are many other examples of eradication which could be quoted and which would illustrate the diversity of the physical methods employed but such detail is for the specialist. Of more interest to our present discussion is the effect of eradication on the progress of an epidemic. We return to Van der Plank's model:

$$x = x_0 e^{rt} \qquad (4.1)$$

By taking logarithms and transforming (as on p. 35) we can derive an expression which tells us the time it takes to reach any disease level (x) starting with out inoculum x_0:

$$t = \frac{1}{r} \log_e \frac{x}{x_0} \qquad (4.2)$$

and we can convert this equation (approximately) to base 10 logarithms by multiplying by 2.3:

$$t = \frac{2.3}{r} \log_{10} \frac{x}{x_0} \qquad (4.3)$$

Now supposing we have two fields. In one (which we will distinguish with the subscript a) there is 0.001 inoculum—it could be that one plant in every thousand is diseased or that there is one resting spore of a pathogenic fungus per thousand m^2 of the surface soil. In the other (subscript b), this inoculum has been reduced to 0.0001 (i.e. a 90% reduction) by some eradication programme, e.g., the physical removal of diseased plants. Conditions are otherwise identical so that the infection rate r is the same in both fields. In this situation the times taken for the disease to reach the same level x, in both fields will be respectively

$$t_a = \frac{2.3}{r} \log_{10} \frac{x}{0.001}$$

$$t_b = \frac{2.3}{r} \log_{10} \frac{x}{0.0001}$$

Clearly t_b is greater than t_a and indeed the difference between them, which is the delay in reaching the same disease level in the epidemic, is a measure of the benefit of the eradication programme. This delay, $t_b - t_a$, we denote by Δt. Thus,

$$\begin{aligned}\Delta t &= \frac{2.3}{r}\left[\log_{10}\frac{x}{0.0001} - \log_{10}\frac{x}{0.001}\right] \qquad (4.4)\\ &= \frac{2.3}{r} \log_{10}\left[\frac{x}{0.0001} \div \frac{x}{0.001}\right]\\ &= \frac{2.3}{r} \log_{10}\left[\frac{0.001}{0.0001}\right]\\ &= \frac{2.3}{r} \log_{10} [10]\\ &= \frac{2.3}{r}\end{aligned}$$

or in general terms, $\Delta t = \frac{2.3}{r} \log_{10} \frac{x_0}{x_{0_s}}$ (4.5)

where x_0 is initial inoculum and x_{0_s} that left after eradication. We are left in equation (**4.4**) above with a simple expression in which the value of r, the infection rate, is all important. So let us substitute some reasonably-realistic figures for it. In a favourable year $r=0.46$ (per unit of disease per day) for potato blight; in a 'non-blight' year it might be only half this. This indicates that if we could reduce initial inoculum by 90% we might expect a delay of 5 days ($\Delta t = 2.3/0.46$) in a 'blight' year and a delay of 10 days in a 'non-blight' year. The benefits are somewhat marginal and in most situations scarcely warrant the effort involved in attempting a 90% reduction of initial inoculum. Removal of infected plants early in the season is certainly not practical. However, in many instances blight starts in fields near piles of potatoes which have been discarded either from stores or from plantings at the end of the previous season. Some of these tubers give rise to blighted sprouts. Removal of these sources of inoculum is relatively easy and well worthwhile.

If we substitute smaller values of r in equation (**4.4**) above we get increased values for t, e.g., when $r=0.1$, $t=23$ days; when $r=0.001$, $t=230$. Clearly, reduction of initial inoculum by sanitation has most effect on diseases with a low infection rate.

We now consider other, somewhat less obvious, ways of reducing inoculum.

4.2.2 Cultivars with race-specific resistance

Man has developed crops by breeding to suit his needs. There are, for example, different kinds of potato, their foliage and growth habits differ and so do the tubers—some mature early, others late, some make good potato crisps, others are most suitable for general household use. We call these different kinds, cultivars. There are also different kinds of the blight fungus, *Phytophthora infestans*, which have arisen through mutation and other processes and have been selected naturally in the development of the potato as a crop. These we call races.

Potato cultivars differ in their reaction to races of *P. infestans*; in some combinations the fungus grows well, in others it does not. In these instances we say that the particular cultivar is respectively susceptible or resistant to the particular race of the fungus. Some cultivars are resistant to certain races but susceptible to others. We call this 'race-specific resistance' but other terms have been used; Van der Plank calls it 'vertical resistance'. Whatever the name the effect, epidemiologically, is similar to sanitation, for as Van der Plank points out, if 99% of the spores reaching a field of potatoes are of races of *P. infestans* to which the cultivar is resistant, the epidemic must build up from the remaining 1%.

This type of resistance has long held the attention of plant breeders not only in respect of potatoes and blight but also in relation to other crops and their respective pathogens. One attraction is the relative ease with which this resistance can be tested in progeny from breeding lines because it is an 'all-or-nothing' response. The source of the resistance is often a wild species related to the host. The Central American species *Solanum demissum* with its genes for resistance to *P. infestans* (R-genes) has been much used in breeding blight-resistant potatoes. In work of this type the overall aim is to incorporate into one commercially-acceptable cultivar genes which will collectively confer resistance to all known races. It appears an attractive method of disease control but in practice the results have frequently been less than satisfactory. The introduction of potato cultivars with R-genes will serve as illustration. When the first of these were developed some 40 years ago they appeared completely resistant to blight. Then as the acreage of these cultivars increased slightly a few blight lesions were noted. New races of the fungus had developed to which these cultivars were susceptible. At first these races were scarce but as the acreage of the R-gene cultivars increased so did the amounts of the races able to attack them. The 'all-or-nothing' type of resistance now became a liability and the popularity of the cultivars only served to make matters worse. This effect of popularity can be illustrated theoretically by use of equation (4.5) above. Let us assume that $r = 0.46$ and that initial inoculum is proportional to the abundance of cultivars with R-genes. If only 1% of the potato crops consists of such cultivars then the delay in the epidemic on these (compared with other cultivars without R-genes) is given by

$$\Delta t = \frac{2.3}{0.46} \log_{10} \frac{100}{0.01} = 20 \text{ days.}$$

If 50% of the potato acreage contain R genes Δt is reduced to about $1\frac{1}{2}$ days. These calculations assume the conditions of a blight year. In a 'non-blight' year (say, $r = 0.23$) the corresponding delays in the epidemics would be twice as long. The irony is that in all these situations the development of the resistant cultivars themselves encourages the evolution of new races. Further, benefits of the resistance are then least when they are most required—when the cultivars are widely grown and when conditions favour the pathogen.

4.2.3 Systemic fungicides

It was long the fungicide manufacturers' dream to produce materials that would be absorbed by plants and become generally distributed within them with no harmful effects, yet able to prevent the entry and growth of pathogenic fungi. In recent years such systemic fungicides have become a reality (Table 3). Some, like benomyl (which is a short name for the chemical, methyl N-[1-(butyl-carbamoyl)-2-benzimidazolyl] carbamate) are effective against many pathogens: we say it has a wide spectrum of activity. Others effectively control only one or two pathogens, like ethirimol (the chemical, 5-butyl-2-ethylamino-4-hydroxy-6-methylpyrimidine) which is used as a seed dressing on barley to limit the subsequent development of powdery mildew within the crop.

At present there are still relatively few of these materials and some scarcely deserve the term systemic because they do not become generally distributed within the plant but are absorbed only into the leaves that are sprayed. None as yet have been found which become generally distributed within woody plants and hardly any that are effective against fungal pathogens belonging to the Phycomycetes which includes *Phytophthora infestans* and the many downy mildews such as *Plasmopara viticola* on the grape-vine. No doubt these deficiencies will be made good in time but already there are indications of a much more serious problem. We have already seen that in many instances the development of cultivars with race-specific resistance leads also to the production of new races of the pathogen able to grow on them. It is now clear that the widespread use of a systemic fungicide also imposes a selection pressure on the pathogen population and that new races can evolve which are able to grow on treated plants. The two situations are similar. Thus there are already reports, amongst others, of benomyl-resistant strains of *Sphaerotheca fuligena*, the powdery mildew fungus of cucurbits, and of ethirimol-resistant strains of *Erysiphe graminis*. Fortunately there is room for manœuvre—the use of systemic fungicides can be alternated with more conventional materials to prevent the build-up of the resistant strains. Similarly, chemically-different systemic fungicides which control the same pathogen may be used successively, provided it is clearly shown that strains able to grow on plants treated with one fungicide do not grow on plants treated with the other.

4.3 Reducing infection rate

4.3.1 Foliar fungicides

Application of foliar fungicides is the conventional method of reducing infection rate. Figure 4–1 shows the effect on blight of spraying potatoes with a copper fungicide. Here disease progress has been plotted as $\log_e [x/1-x]$ against time where x is the proportion of diseased tissue. This is a device which gives a visual picture of infection rate r, for it is the slope of the line which is fitted mathematically to the points obtained. Clearly in this example only partial control was obtained and this is the usual pattern. This is simply because it is virtually impossible to obtain a complete covering of the chemical on the plant which would thus inhibit all spores and also the deposits which are obtained become eroded by

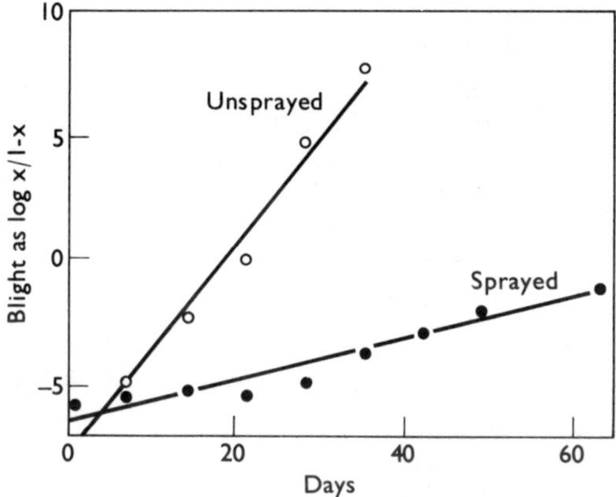

Fig. 4–1 The effect on spraying with a copper fungicide on the rate of infection of potato blight. (Based on VAN DER PLANK, J. E. (1963).)

wind and rain. The degree of control is seldom so dramatic as with the systemic fungicides nor generally are the materials so sophisticated. Perhaps partly because of this races of pathogens which tolerate these materials are seldom a problem despite many years of usage. The development of a strain of the fungus *Pyrenophora avenae* tolerant to mercurial seed dressings on oats remains an almost unique exception. There has, of course, been some sophistication in the development of these chemical protectants since the middle of the nineteenth century when simple compounds like copper sulphate and mercuric chloride were in vogue and the best fungicide was undoubtedly a mixture of copper sulphate and lime called Bordeaux mixture. Table 4 lists some of

Table 3 Systemic fungicides

Group	Examples*
1 Organophosphorus compounds	O,O-di-isopropyl-S-benzyl thiophosphate ('Kitazin')
2 Carboxylic acid anilides	2,3-dihydro-6-methyl-5-phenylcarbamoyl-1,4-oxathiin (carboxin)
3 Heterocyclic compounds	
(a) Benzimidazoles	Methyl N-[1-(butylcarbamoyl)-2-benzimidazole] carbamate (benomyl)
(b) Hydroxypyrimides	5-n-butyl-2-ethylamino-4-hydroxy-6-methylpyrimidine (ethirimol)
(c) Dimethyl morpholins	N-tridecyl-2,6-dimethylmorpholine (tridemorph)*
4 Aromatic compounds	1,2-di-(3-ethoxycarbonyl-2-thioureido) benzene (thiophanate)

* There are often three names used to describe a fungicide: (1) the chemical name of the active ingredient, e.g., N-tridecyl-2,6-dimethylmorpholine (2) a common (or 'coined') name for the active ingredient to avoid use of the often long chemical name; this coined name is usually first approved by official bodies such as the British Standards Institution (BSI) or International Organization for Standardization (ISO), e.g., tridemorph and (3) a trade name used by the manufacturer for a formulation of this chemical, e.g., a 75% emulsifiable concentrate of tridemorph is manufactured by the company, Badische Anilin- & Soda-Fabrik AG, under the trade name 'Calixin'. (Information based on *Pesticide Manual*, 3rd edition (1972), edited by H. MARTIN and issued by the British Crop Protection Council and on *Systemic Fungicides* (1972), edited by R. W. MARSH, Longman, London.)

the more modern materials. Correspondingly there have been marked improvements in applying fungicides not only in the design of machinery but also in timing of spray applications as the next chapter indicates.

4.3.2 Cultivars with race non-specific resistance

The failure of potato cultivars with resistance based on the R-genes has led breeders to look at other types of resistance. There are, for example, other cultivars which react similarly to all races and show some general level of resistance. This is called race non-specific resistance or in Van der Plank's terminology, horizontal resistance. It is quantitative in effect: on such cultivars the number of infections is reduced, the fungus takes longer to sporulate and its rate of reproduction is reduced. The net result

Table 4 Some modern foliar fungicides

Group	*Examples*
1 Copper	Several inorganic compounds such as copper oxide and copper oxychloride are used, either alone or mixed with the dithiocarbamate, maneb
2 Sulphur	Various finely-divided preparations of elemental Sulphur are still used as dusts or wettable powders from which aqueous suspensions can be prepared
3 Mercurial	Phenyl mercury acetate
4 Organotin compounds	triphenyl tin acetate (fentin acetate)
5 Dithiocarbamates	Manganese ethylene bio-dithiocarbamate (maneb)
6 Quinones	2,3-dichloro-1,4-napthoquinone (dichlone)
7 Dinitro-compounds	A mixture of 2,4-dinitro-6-octylphenyl crotonate (1) and 2,6-dinitro-4-octophenyl crotonate (2) known as dinocap
8 Miscellaneous compounds	N-(trichloromethylthio) cyclohex-4-en-1,2-dicarboximide (captan) N-(1,1,2,2-tetrachloroethylthio) cyclohex-4-en 1,2-dicarboximide (captafol)

Information based on *Pesticide Manual*, 3rd edition, 1972. Edited by A. MARTIN and issued by the British Crop Protection Council.

is that the epidemic is slowed down, i.e. *r* is reduced. Cultivars with this type of resistance have one outstanding advantage: they react similarly to any new pathogenic race which arises by mutation or other processes and so do not succumb dramatically as do those with race-specific resistance. Unfortunately the quantitative nature of race non-specific resistance, based as it is on a complex of relatively-minor morphological and physiological characteristics, makes its determination difficult in progeny within breeding lines. It is for this reason that it remained so long neglected by breeders.

The effects of race non-specific resistance on the progress of an epidemic are discussed fully by VAN DER PLANK (1963). One of the more interesting aspects is his clear demonstration that fungicidal activity can be enhanced by the incorporation of race non-specific resistance into cultivars. The details of his analysis are beyond the scope of this account, suffice it to say that a continuous production of a certain level of spores is necessary for an epidemic to continue and this can be determined for a particular disease if certain parameters are known. Clearly the aim of applying fungicides is to destroy spores to a level insufficient to maintain the epidemic. Table 5 shows the destruction of spores required to stop an

Table 5 The calculated percentage of spores that a fungicide must destroy in order to stop an epidemic of potato blight (modified from VAN DER PLANK (1963), *Plant Diseases: Epidemics and Control*. Academic Press)

Infection rate (per unit per day)	*% destruction of spores needed*
0.5	89
0.4	83
0.3	74
0.2	59
0.1	36

epidemic of potato blight in situations which result in different infection rates. From this we see that approximately 8 spores out of 9 need to be killed when $r=0.5$ but only 1 spore out of 3 when $r=0.1$. Lowering infection rate (e.g. by introducing race non-specific resistance) clearly benefits fungicidal control.

This leads us to two final comments. One disease is seldom controlled in a crop by one method alone; potato cultivars are developed with resistance to blight and crops are sprayed. There is also seldom one disease of a particular crop, so some integration of the various control measures is required to maintain its health.

5 Predictions about Epidemics: Methods

The timing of control methods is often critical if an epidemic is to be halted. In recent years various techniques have been evolved for forecasting the onset of an epidemic, for estimating its severity or progress and for taking decisions in relation to control. These are indicated in this chapter.

5.1 Simple empirical relationships

Investigations at the beginning of this century indicated the importance of moisture and temperature in relation to both the sporulation of *Phytophthora infestans* and its ability to invade potato leaves and it was soon appreciated that blight outbreaks were favoured by cool, moist weather. With an increasing realization that sprays had to be applied early if the disease was to be controlled there were attempts to determine more accurately the relationships between blight and the weather so that the first outbreaks could be predicted.

Workers in the Netherlands were amongst the first to define the weather conditions necessary for the outbreak of blight in a potato crop and this was further developed in England by Beaumont. He examined critically the meteorological data for the periods preceding blight outbreaks in SW. England and from this deduced that if for 2 days there was a minimum temperature of 10°C with relative humidity of 75% or over, then blight could be expected approximately 10 days afterwards. These periods are called 'Beaumont periods' and are now used in England as a basis of a forecasting system. Slightly different criteria are used in other countries but the aim is basically similar, to characterize conditions which allow germination of the fungal spores and subsequent penetration of the leaf. In all of these some allowance is made for the growth of the crop—there must obviously be an abundance of susceptible tissue—but inoculum is considered only in very general terms. However, some appraisal of inoculum is often an important feature of forecasts of this type for other diseases, e.g., apple scab.

In many areas where apples are grown the fungus, *Venturia inaequalis*, overwinters in the fallen leaves and in spring spores (ascospores) are discharged from the flask shaped structures (pseudothecia) which develop in these leaves. These ascospores are then blown by wind to the young leaves and developing fruit where they cause unsightly scabbed areas. Temperature and the duration of leaf wetness mainly determine infection by these ascospores and as early as 1954 two workers at Cornell University in the United States compiled tables which indicated the weather conditions leading to light, moderate and severe infections based on their observations over many years. For forecasts of apple scab outbreaks the minimum conditions, that is conditions for light infection, are appropriate and some examples are listed in Table 6. Each temperature/leaf wetness combination (e.g. 10°/14 h) is now known as a 'Mills period' and can be used to time applications of sprays. These periods can be assessed from temperature data and those obtained by a 'surface-wetness' recorder, a device in which the deposition of rain or dew and its subsequent evaporation is measured from changes in weight of a polystyrene block. Alternatively predictions are based on 'Smith periods' in which a corresponding period of relative humidity of 90% or over replaces leaf wetness and thus the need for a special recorder. Several countries now have warning systems based on these criteria and an appraisal of ascospore development and discharge. For example, in the Netherlands a careful watch is kept on the development of the fungus by the Plant Protection service by means of laboratory tests on overwintered leaves. Weather conditions are assessed by a network of stations and all this information is fed back to a main centre from which warnings are issued when (a) ascospores are ready to mature, (b) an ascospore release is expected, (c) an ascospore release is reported, (d) infection periods have occurred and (e) ascospore discharge has ceased.

Table 6 Some examples of 'Mills periods'

Mean temperature (°C) over period	5.6	7.2	10	11.7	15
Hours of leaf wetness	30	20	14	11	10

5.2 Regression analysis

Often an indication that conditions are favourable for the outbreak of a disease is not enough. An answer to the question 'How much disease?' is required. Even the early work on apple scab attempted some general assessment of this kind. The techniques of regression analysis have been much used in such predictions. In simple terms an attempt is made to relate different amounts of disease to different quantities or values of some parameters which are readily measured. This analysis is either based on some data which already exist or experiments are initiated to examine likely relationships and provide the necessary data. The variability of any relationship can be determined by standard statistical methods and an equation derived which may then be used predictively.

5.2.1 *Linear regression*

In the simplest situation disease is related linearly to one parameter. This relationship is expressed generally by the equation $y=a+bx$, where y=amount of disease, x is the measured parameter and a and b are constants. In statistical language we are concerned with the regression of y (called the dependent variable) on x (called the independent variable). We take as an example predictions about the severity of club-root of crucifers. The causal fungus of club-root, *Plasmodiophora brassicae*, perennates in soil as resting spores, and invades young cabbage, cauliflower or other cruciferous seedlings which are planted in contaminated sites. The invaded roots become swollen and 'clubbed' and later in the season cracks and fissures develop as a result of this uneven growth through which other micro-organisms then enter. The clubs decay and more resting spores of the fungus are thus returned to the soil. Epidemiologically there is only one 'generation' of the fungus during the season and the amount of disease in the crop depends essentially on the amount of initial inoculum. In the U.K. forecasts of the likely incidence of club-root are based on assessments of this inoculum before planting. Soil samples are collected at about 100 points in fields of 3 acres (*c.* 2.5 ha) or less, the soil is mixed, passed through a ¼ in (6.3 mm) sieve, fertilizer is added and the mix is put into 5 in (12.7 cm) pots. Each pot is then planted with five rape seedlings (*Brassica napus*) grown in sterilized compost for 10 days. The pots are placed in a greenhouse at 21° C and the soil maintained at about 70% of its water holding capacity. Plants are examined in 5 to 6 weeks for infection which is usually evident at this stage as one large club on the tap root. Results are expressed as percentage of plants showing infection. The relationship between these figures and the incidence of club-root subsequently observed in the sampled fields is shown in Fig. 5–1.

5.2.2 *Partial regression*

It is not always possible to relate disease to one factor. As an example we shall consider the disease of sugar beet called virus yellows in which there is a yellowing and crinkling of the leaves and a subsequent loss of sugar to the roots. In the U.K. there are actually two viruses associated with yellows—beet yellows virus and beet mild yellows virus, but our discussion relates primarily to the former. Beet yellows virus is transmitted by aphids and predominantly by the green peach aphid, *Myzus persicae*, so the availability and behaviour of this aphid markedly affects disease development. The virus itself overwinters mainly in stored beet and the aphid acquires the virus either from leaves which sprout from stored, infected roots or from infected plants of beet grown for seed and which persist through the winter. In mild winters many aphids survive, some also on beet in store or in seed beds, others on crops like

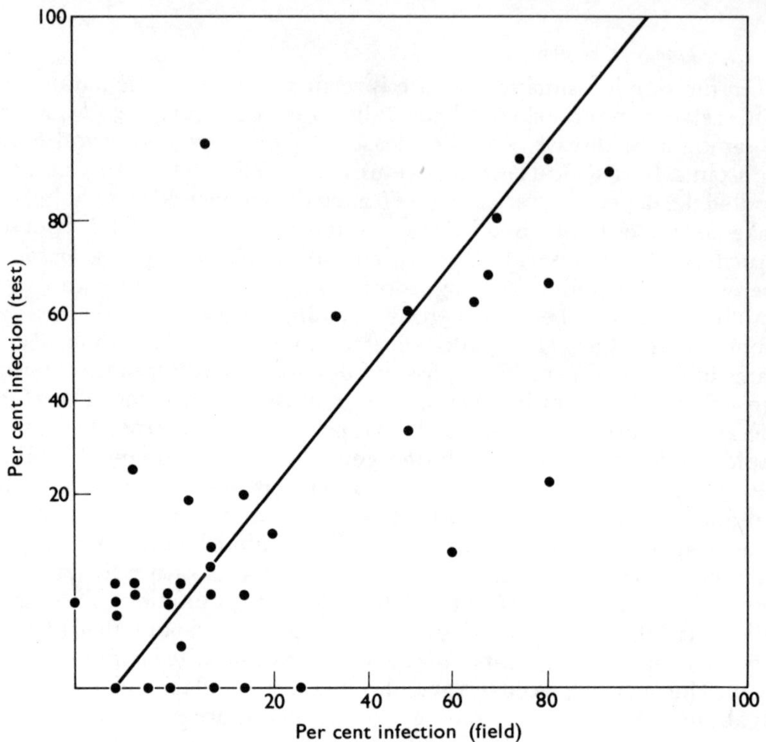

Fig. 5–1 Relationship between laboratory test results and corresponding field assessments of club-root. (Based on MELVILLE, S. C. and HAWKEN, R. H., (1967). *Plant Pathology*, **16**, 146. By permission of the Controller, H.M.S.O.)

cabbages; in severe winters relatively few aphids survive as eggs on peach. In both instances aphid multiplication and spread is then determined by the spring weather, dry warm weather and calm conditions being the most favourable. So, generally, severe aphid infestations and therefore severe outbreaks of virus yellows in beet are associated with mild winters and warm springs. Now, if we wish to relate beet yellows to these factors we need an equation involving two parameters, i.e. $y = a + b_1x_1 + b_2x_2$ where y is an estimate of beet yellows, x_1 is some measure of the severity of the winter and x_2 is some measure of the warmth of the spring. The figures b_1 and b_2 are called partial regression coefficients—hence the title of this section. Such an analysis has been carried out using the data from previous years. These data were transformed in various ways to meet certain mathematical requirements but these need not concern us. Suffice it to say that y was based on the percentage of plants found to be infected with virus yellows at the end of August, x_1 on the number of freezing days in January, February and March and x_2 on the mean weekly temperature in April. That the derived equation described well the relationship between these factors is shown in Fig. 5–2 where the values calculated

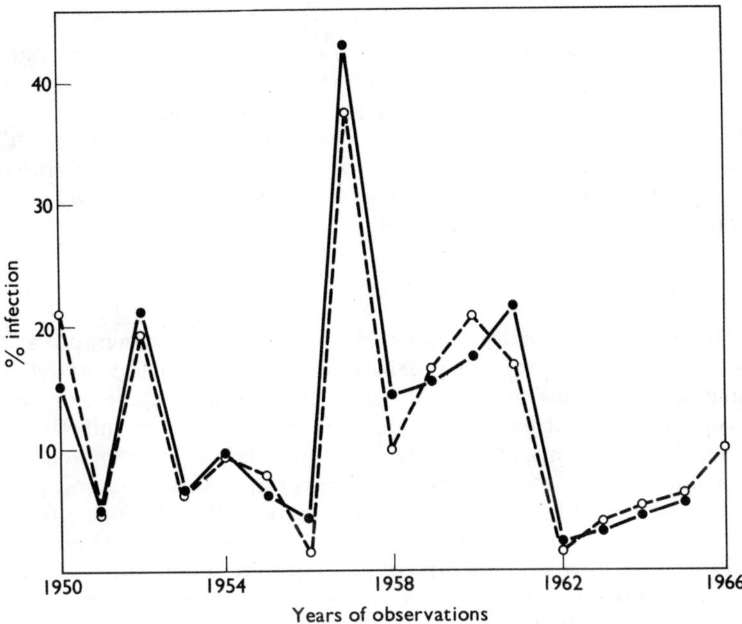

Fig. 5–2 Observed values (●) for % beet yellows compared with values calculated (o) from the regression of % yellows on number of freezing days in January to March (x_1) and deviations from average April temperatures (x_2). (From WATSON, M. A. (1966). *Plant Pathology*, **15**, 147. By permission of the Controller, H.M.S.O.)

from the regression are compared with those actually observed in the field. Because there is this close agreement it ought to be possible to assess by the end of April the likely disease situation in any year and the need for control.

5.2.3 *Multiple regression*

With most diseases caused by air-borne pathogens the relationship between the weather and disease intensity is complex. The weather influences the number of spores which a pathogenic fungus produces within a lesion, the rate at which these spores mature and are dispersed, it affects their deposition on other hosts, their subsequent germination and penetration and to a certain extent the rate at which new lesions develop. The amount of disease at any one time, therefore, may depend not only on the amount present some 10 days or so earlier but also on the number of spores that became airborne, and the subsequent climatic conditions such as rainfall, humidity and temperature. In these situations to analyse relevant field data we now need to use the techniques of multiple regression and derive an equation of the form:

$$y = a + b_1x_1 + b_2x_2 \ldots b_nx_n$$

where y (the dependent variable) is some measure of the amount of disease and $x_1, x_2 \ldots x_n$ (the independent variables) are measures of these different factors. For some diseases, such as potato blight or leaf rust of wheat, there is already a mass of data which has facilitated analyses of this type. So far the predictive value of the derived equations is limited but the attempts to quantify the effects have not infrequently shed new light on the factors influencing the epidemics.

5·3 Simulators

The regression equations indicated in the preceding paragraph can be used as models for predicting the amounts of particular diseases. These equations are, however, based essentially on past epidemics. They illustrate in general how climatic conditions are likely to influence the overall development of the epidemic as indicated by past experience but the components of the epidemic—sporulation, dissemination of spores, etc.—cannot readily be identified and there is thus no place for the incorporation of data from controlled laboratory experiments. In contrast these form the bases for computer programmes which attempt to simulate epidemics. The essential feature of these simulators is that in each there is a logical sequence of steps which corresponds to the components of an epidemic. Simulators have been devised now for several diseases. We shall outline the general principles involved by referring to EPIMAY a simulator of southern corn leaf blight produced by researchers at the Connecticut Agricultural Experiment Station. This disease, caused by the fungus *Helminthosporium maydis*, has been known for some time in the U.S.A. but was generally considered unimportant because the hybrid maize generally planted was resistant to it. But by 1968 there were indications of a change in the pathogen and in the following year it was confirmed that there was a new race of the fungus able to grow on the hitherto-resistant cultivars. A widespread and destructive epidemic followed in 1970.

Our consideration of the simulator can conveniently start with a standard life cycle for a fungus of this type (Fig. 5–3) and then proceed to a more detailed cycle (Fig. 5–4) which incorporates observations specific to *H. maydis*. The fungus produces its conidiophores (stalks) in the lesions. These stalks may be freshly formed (green) or they may have dried since their formation (dry) and this is an important distinction because they become spore-producing (fertile stalks) at different rates. It is also possible, of course, for a fertile stalk to dry out, hence fertile stalks → dry

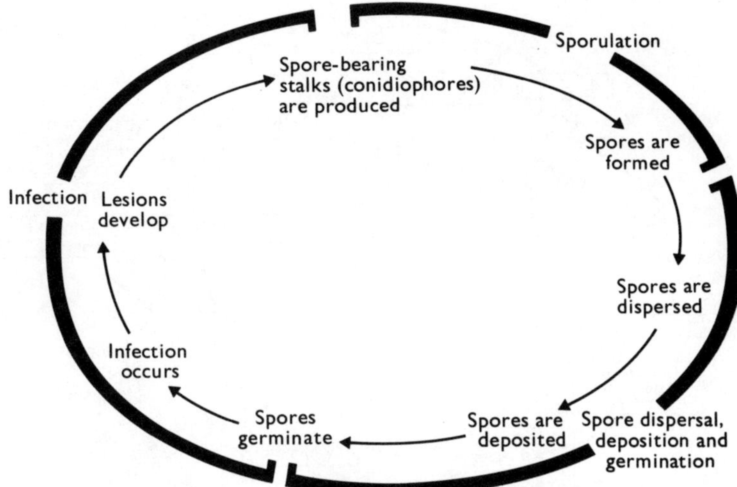

Fig. 5–3 Typical infection cycle of a leaf-infecting fungus.

stalks. Spores are carried to new leaves in water or air but they germinate effectively only in water hence the insertion of a component 'wet caught spores' which comprises water-borne spores plus wind-blown spores that become wet. We next make three changes to arrive at the flow diagram of Fig. 5–5. There are now both solid and dotted lines. The solid lines imply flow of material, e.g., green stalks become dry stalks through the action of

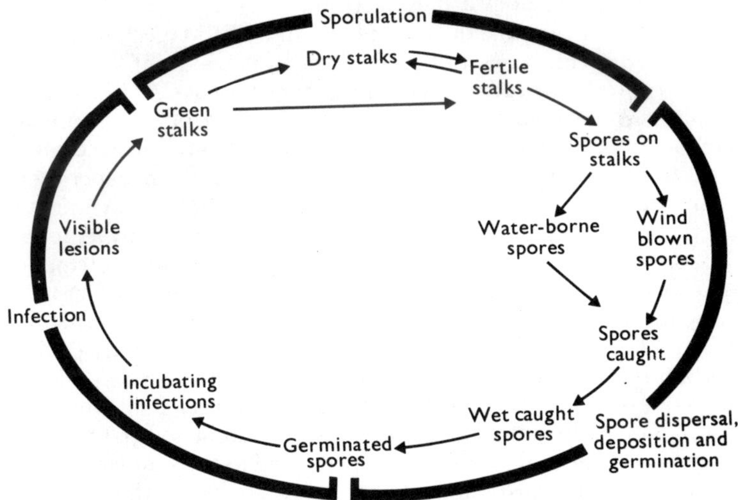

Fig. 5–4 Infection cycle of the fungus, *Helminthosporium maydis*, on maize.

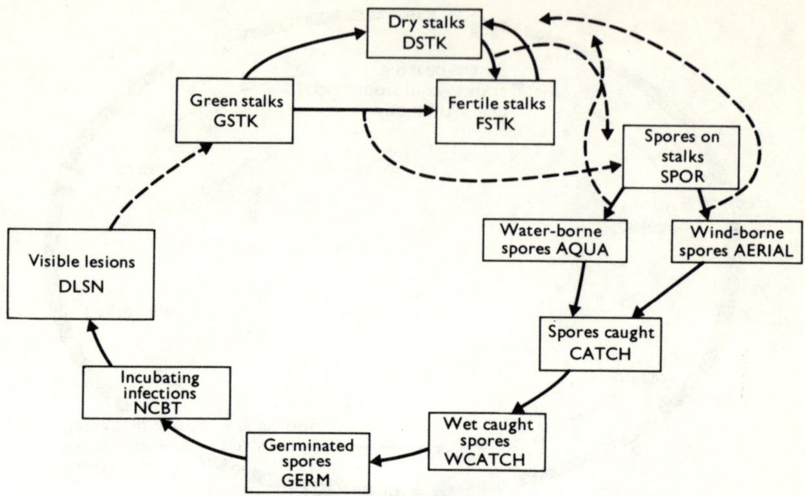

Fig. 5–5 Flow diagram: infection cycle of *Helminthosporium maydis*. (From WAGGONER, P. E., HORSFALL, J. G., and LUKENS, R. J. (1972). *Bulletin of the Connecticut Agricultural Experiment Station, New Haven*, no. 729.)

drying, the mycelium in the lesion results from growth of the spore. In contrast the dotted lines indicate influence or flow information, e.g., lesions influence the production of stalks but they are not stalks. Also, each stage is now given a symbol, e.g., GSTK for green stalks and each is enclosed in a box. These boxes can be considered as accounts, the levels of which fluctuate as does a bank account when money is put into it or withdrawn. For example the level of account GERM (germinated spores) is increased by movement of material from the account WCATCH (wet caught spores) and correspondingly the level of WCATCH is lowered. Simply, some wetted spores germinate. To construct the simulator we need quantitative information about the flow from one account to the next. It is this information which is obtained from experiments in controlled environments. Thus given that spores are wet we might guess that the course of germination will be largely influenced by temperature. This proved to be so and three simplified germination/temperature curves are shown in Fig. 5–6. There may be several climatic factors which act on the flow from one account to the next and all this information has to be incorporated into the programme. Indeed each part of the flow diagram shown in Fig. 5–5 requires amplification. Such detail is beyond the scope of this discussion. The main point is that when all the information is included we have in effect a simulated pathogen. Field observations about the weather are presented to the simulated pathogen by the computer programme and its reactions to this weather are calculated. The sequence of events are summarized and we get in response

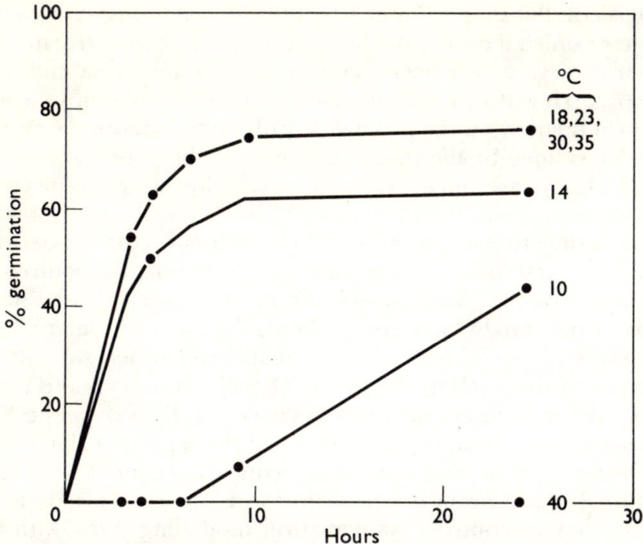

Fig. 5–6 The course of germination for 'wet caught' spores of *Helminthosporium maydis* at various temperatures (°C). (From WAGGONER, P. E., HORSFALL, J. G. and LUKENS, R. J. (1972). *Bulletin of the Connecticut Agricultural Experiment Station, New Haven,* no. 729.)

to this feeding-in of weather data an index of how the epidemic will react to it. We have a forecast.

Clearly the compilation of a simulator is a considerable task but the adaptability of a simulator—to changes in virulence of the pathogen or to the introduction of a more resistant host—make it a powerful tool in epidemiological studies.

5·4 Decision theory

Consider a farmer in the U.K. or N. Europe who grows barley. He knows that in some years attacks of powdery mildew on this crop are severe, equally he now knows that a seed dressing of ethirimol will control this disease for much of the season. But this seed dressing costs money. Will it be worth it? He has to make a decision at planting before he can see the amount of disease. There are similar uncertainties with many methods of plant disease control, especially where pesticides are applied. It's a chancy business, farming. But there are mathematical models which deal with problems involving uncertainty and one of the most recent innovations in studies of the economics of crop disease control has been the introduction of such models, based on Bayesian decision theory. The details of this approach are quite beyond the level of this book but the concept is not. Put simply, it is to provide a formal framework for making decisions in which the costs and benefits of applying a control method for a particular disease have been weighted against the probabilities of that disease occurring at various intensities. In these calculations the effect of the control method on the profitability of the crop looms large and this will be determined by the cost of the control, the potential loss in yield caused by the disease and the reduction in this loss which results from the control method itself.

Let us assume that a control for a fungus disease costs £6 per hectare. There are two actions the farmer can take: he applies the control (action a_1) or not (action a_2). Now diseases, if they occur, vary in intensity. In his crop the farmer might have eventually no disease, or little, moderate or severe levels. These different 'states of nature', which we will denote respectively by the symbols, Θ_1, Θ_2, Θ_3, Θ_4 will result in various monetary losses. To develop the argument let us say £0, £1, £8, £20 per hectare. So, if the disease is severe, applying the control could prevent a potential loss of £20/ha which, if we now subtract the cost of the control itself, means a benefit of £14 per hectare. Calculations of this type are basic to any considerations of control; mathematical modelling starts with further thoughts about the distribution of these 'states of nature' which we can only indicate here. Table 7 shows for our example the outcome of applying the control or not for the four disease situations. It also shows

Table 7 Outcomes (in £'s per hectare) of applying a control or not in different situations (see text for further explanation)

Disease situation	*No control applied (action a_1)*	*Control applied (action a_2)*
No disease (Θ_1)	0	−6
Slight infection (Θ_2)	−1	−5
Moderate infection (Θ_3)	−8	+2
Severe infection (Θ_4)	−20	+14
Average outcome	−4.75	+1.25

(by summing the columns and dividing by four) the average outcome for each action. This suggests that it pays to apply the control whatever the disease level. However, as an average it assumes that the chances of each disease level occurring are equal. Supposing they are not? There could be some past records which might suggest otherwise. If, for example, in the last 10 years there were two light outbreaks, two moderate, five severe and one year without disease we should have the following probabilities (P)

$$P(\Theta_1)=0.1 \text{ (i.e. 1 year in 10)}$$
$$P(\Theta_2)=0.2$$
$$P(\Theta_3)=0.2$$
$$P(\Theta_4)=0.5$$

This information can now be used to 'weight' the outcomes shown in Table 7. The overall expected value of action a_1 (no control) is calculated as:

$$(0)(0.1)+(-1)(0.2)+(-8)(0.2)+(-20)(0.5)=-11.8$$

i.e. a loss per hectare of £11.8.

And of a_2 (control) similarly:

$$(-6)(0.1)+(-5)(0.2)+(+2)(0.2)+(+14)(0.5)=+5.8$$

i.e. a gain per hectare of £5.8.

In this very simple example only the margin of profit changes (from applying the control) but in other, more realistic situations where several courses of action are possible (e.g. seed dressing and/or 1, 2, 3 or more sprays) even calculations to this stage can be very revealing. The model is then developed by incorporating further information about the probable disease level using a forecast, e.g., one which relates disease level to winter or spring temperatures like that described above for beet yellows.

The use of such models, as with those that simulate epidemics, is as yet only beginning. Common sense tells us not to expect too much of predictions about plant diseases for if we look at our 'disease triangle (p. 3) we see that we are in part making predictions about the environment. Nevertheless, the future prospects are exciting.

Further Reading

AGRIOS, G. N. (1969). *Plant Pathology*. Academic Press, New York and London.

CARTER, W. (1973). *Insects in Relation to Plant Disease*. John Wiley, New York.

DEVERALL, B. J. (1969). *Fungal Parasitism*. Edward Arnold, London.

EVANS, E. (1968). *Plant Diseases and their Chemical Control*. Blackwell Scientific Publications, Oxford.

GOODMAN, R. N., KIRALY, Z. and ZEITLIN, M. (1967). *The Biochemistry and Physiology of Infectious Plant Disease*. Van Nostrand, Princeton.

LARGE, E. C. (1940). *The Advance of the Fungi*. Jonathan Cape, London.

ROBERTS, D. A. and BOOTHROYD C. W. (1972). *Fundamentals of Plant Pathology*. Freeman, San Francisco.

TARR, S. A. J. (1972). *Principles of Plant Pathology*. Macmillan, London.

VAN DER PLANK, J. E. (1963). *Plant Diseases: Epidemics and Control*. Academic Press, New York and London.

WHEELER, B. E. J. (1969). *An Introduction of Plant Diseases*. John Wiley, London.

WHEELER, B. E. J. (1968). Fungal parasites of plants. In *The Fungi*, 3, 179–210. Academic Press, New York and London.

WOOD, R. K. S. (1967). *Physiological Plant Pathology*. Blackwell Scientific Publications, Oxford.

WOODHAM-SMITH, C. (1963). *The Great Hunger, Ireland, 1854–9*. Hamish Hamilton, London.

Additionally there are many useful reviews of particular aspects of plant pathology in the *Annual Review of Phytopathology*, published by Annual Reviews, Inc., Palo Alto, California.